# Springer Tracts in Modern Physics 79

Ergebnisse der exakten Naturwissenschaften

*Manuscripts for publication should be addressed to:*

Gerhard Höhler
Institut für Theoretische Kernphysik der Universität Karlsruhe
Postfach 6380, D-7500 Karlsruhe 1

*Proofs and all correspondence concerning papers in the process of publication should be addressed to:*

Ernst A. Niekisch
Institut für Grenzflächenforschung und Vakuumphysik
der Kernforschungsanlage Jülich, Postfach 1913, D-5170 Jülich

# Elementary Particle Physics

Contributions by
E. Paul H. Rollnik P. Stichel

With 37 Figures

Springer-Verlag
Berlin Heidelberg New York 1976

Dr. Ewald Paul

Physikalisches Institut der Universität Bonn, Nussallee 12, D-5300 Bonn 1

Professor Dr. Horst Rollnik

Physikalisches Institut der Universität Bonn, Nussallee 12, D-5300 Bonn 1

Professor Dr. Peter Stichel

Fakultät für Physik der Universität Bielefeld, D-4800 Bielefeld



ISBN 3-540-07778-2 Springer-Verlag Berlin Heidelberg New York
ISBN 0-387-07778-2 Springer-Verlag New York Heidelberg Berlin

Library of Congress Cataloging in Publication Data. Paul, Ewald, 1937 —. Elemantary Particle Physics (Springer tracts in modern physics; 79). Bibliography: p. Includes index. 1. Weak interactions (Nuclear physics). 2. Mesons-Decay. 3. Compton effect. I. Rollnik, Horst, 1931 — II. Stichel, P. III. Title. VI. Series. QC1. S797. vol. 79. [QC794.8. W4]. 539'.08s. [539.7'54]. 76—14422.



Offset printing and bookbinding: Brühlsche Universitätsdruckerei, Giessen.

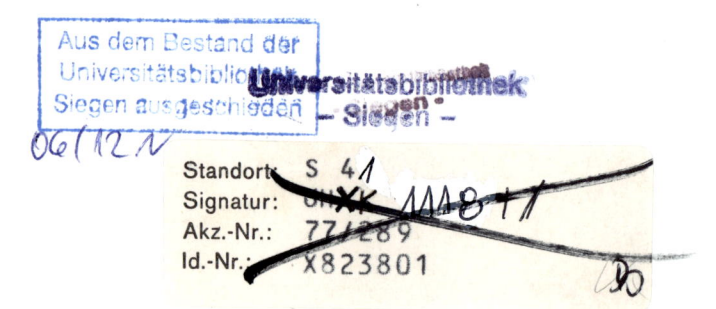

# Contents

## Compton Scattering

By *H. Rollnik* and *P. Stichel*. With 16 Figures



## Status of Interference Experiments with Neutral Kaons

By *E. Paul*. With 21 Figures

# Compton Scattering

H. Rollnik and P. Stichel

## 1. Compton Scattering in the Resonance Region

### 1.1 Forward Scattering: Low-Energy Theorems, Dispersion Relations and Sum Rules

The rôle of the elastic scattering of light on elementary particles in the course of the formulation and clarification of the laws which govern the world of particles and especially their electromagnetic interaction cannot be overestimated. Already before the advent of quantum theory, Thomson scattering, i.e., the elastic scattering of light on free electrons, played an important part in atomic physics. In classical electrodynamics the total cross section for this process is energy independent and given by the Thomson cross section [1]

$$\sigma_{\text{Thomson}} = \frac{8\pi}{3}\left(\frac{e^2}{m_e}\right)^2 \qquad e^2 = \alpha = \frac{1}{137} ; \qquad m_e = \text{electron mass} . \tag{1}$$

Each textbook on quantum theory describes how the experiments of Compton on photon-electron scattering clarify the basic quantum properties of light.

Within modern particle physics the first important achievement in the theory of Compton scattering was the proof of a low-energy theorem by THIRRING in 1950 [2]. According to this theorem the Thomson formula (1) is exactly valid at threshold--to any order in the electric charge--if e and $m_e$ are interpreted as the renormalized charge and electron mass, respectively. A generalization of this result was obtained in 1954 by LOW et al. [3]. To describe this result we consider forward Compton scattering. In this simple kinematical situation the scattering amplitude can be written in the Center of Mass System (CMS) as

$$T^{\text{CMS}} = \chi_f^+ [f_1(\nu)(\underset{\sim}{\varepsilon}' \cdot \underset{\sim}{\varepsilon}) + f_2(\nu) i \underset{\sim}{\sigma}(\underset{\sim}{\varepsilon}' \times \underset{\sim}{\varepsilon})] \chi_i . \tag{2}$$

Here $X_i$ and $X_f$ are Pauli spinors describing the electron spin before and after the scattering; $\underset{\sim}{\varepsilon}$ and $\underset{\sim}{\varepsilon}'$ are the polarization vectors of the incident and final photon in the Coulomb gauge ($\underset{\sim}{\varepsilon}$ and $\underset{\sim}{\varepsilon}'$ are perpendicular to the process direction). Finally the photon energy $\nu$ in the laboratory system is used

$$\nu = \frac{W^2 - m_e^2}{2m_e} , \quad W = \text{total CMS energy} .$$

Eq.(2) simply follows from the fact that $T^{CMS}$ must be bilinear in $\underset{\sim}{\varepsilon}$ and $\underset{\sim}{\varepsilon}'$ and invariant with respect to spatial rotation and reflection. In terms of $f_1$ and $f_2$ the low-energy theorems read

$$\lim_{\nu\to 0} f_1(\nu) = f_1(0) = -\frac{\alpha}{m_e} , \quad \lim_{\nu\to 0} \frac{f_2(\nu)}{\nu} = f_2'(0) = -\frac{\alpha\kappa^2}{2m_e^2} \tag{3a,b}$$

where $\alpha = e^2$ and $\mu_e = (e/2m_e)(1+\kappa)$ is the full magnetic moment of the electron.

The proof of (3a,b) is based only on gauge-invariance considerations. Therefore it can be directly used also for the Compton scattering off strongly interacting particles, especially off protons and neutrons. Of course, one has to use the appropriate values for the mass, charge and the anomalous moment $\kappa$. In this case (3a,b) are valid in any order of the strong coupling. Thus these relations contain highly non-trivial statements on the physics of strong interactions. They can be considered as the first terms of a series development in powers of $\nu$. From crossing symmetry (see Sec.1.4) one can deduce that $f_1$ is an even, $f_2$ an odd function of $\nu$

$$f_1(-\nu) = f_1(\nu) , \quad f_2(-\nu) = -f_2(\nu) . \tag{4}$$

Thus a power development looks like

$$f_1(\nu) = a_0 + a_1\nu^2 + \cdots , \quad f_2(\nu) = b_0\nu + b_1\nu^3 + \cdots . \tag{5}$$

From (3a,b)

$$a_0 = -\frac{\alpha}{m} , \quad b_0 = -\frac{\alpha\kappa_p}{2m} \tag{5a}$$

where we have now used the proton parameter $m = 936$ MeV, $\kappa_p = 1.79$. The first coefficients of (5) are thus given by the electrostatic properties of the proton. Also the coefficient $a_1$ can be interpreted by a classical electromagnetic property of a particle: $a_1$ is given by the sum of the electric and magnetic polarizability of the proton $\alpha_p + \beta_p$ [4,5]

$$a_1 = \alpha_p + \beta_p \; . \tag{5b}$$

These polarizabilities give rise to the classical phenomenon of Rayleigh scattering. They can be determined experimentally by measuring the differential cross section for very low-energy photons. Recent measurements led to the result [6]

$$\alpha_p^{exp} = (10.7 \pm 1.1) \times 10^{-43}\ \text{cm}^3 \; , \qquad \beta_p^{exp} = (-0.7 \pm 1.6) \times 10^{-43}\ \text{cm}^3 \; .$$

On the other hand, $\alpha + \beta$ can be obtained indirectly by means of forward-scattering dispersion relations.

In fact it was forward Compton scattering where analyticity properties were studied in detail for the first time in the classical paper by GELL-MANN, GOLDBERGER and THIRRING from 1954 [7]. Using causality arguments it was shown that $f_1$ and $f_2$ are holomorphic functions in the complex $\nu$-plane up to cuts from $-\infty$ to $-\nu_0$ and from $\nu_0$ to $+\infty$ with

$$\nu_0 = \text{threshold energy for single pion production}$$

$$= \frac{(m + m_\pi)^2 - m^2}{2m} = m_\pi\left(1 + \frac{m_\pi}{2m}\right) \; .$$

Using Cauchy's theorem and assuming a sufficiently rapid decrease of $f_1$, $f_2$ for large $\nu$-values one finds for the even function $f_1$

$$\text{Re}\{f_1(\nu)\} = \frac{2}{\pi} P \int_{\nu_0}^{\infty} \frac{\text{Im}\{f_1(\nu')\}}{\nu'^2 - \nu^2} \nu' d\nu' \tag{6}$$

and for the odd function $f_2$

$$\text{Re}\{f_2(\nu)\} = \frac{2\nu}{\pi} P \int_{\nu_0}^{\infty} \frac{\text{Im}\{f_2(\nu')\}}{\nu'^2 - \nu^2} d\nu' \; . \tag{7}$$

To analyze the convergence of the integrals we need some kinematical relations [8]. From (2) we find for the scattering amplitude $f_{3/2}$ for photons of positive helicity $\lambda = +1$ described by $\hat{\varepsilon}_+ = -2^{-1/2}(-\hat{e}_1 + i\hat{e}_2)$ (where the incident photon moves along $\hat{e}_3$) on nucleons with parallel spin (helicity $\nu = -1/2$ !)

$$f_{3/2}(\nu) = f_1(\nu) - f_2(\nu) . \tag{8a}$$

The index 3/2 points to the total helicity in photon direction $\lambda - \nu$. Analogously if the spins are antiparallel ($\lambda = +1$, $\nu = +1/2$) with total helicity 1/2

$$f_{1/2}(\nu) = f_1(\nu) + f_2(\nu) . \tag{8b}$$

The normalization is chosen such that the differential cross section in the CMS is given by

$$\frac{d\sigma_{3/2,1/2}}{d\Omega} = \left|\frac{m}{W} f_{3/2,1/2}\right|^2 . \tag{9}$$

(The factor $1/W = (2m\nu + m^2)^{-1/2}$ guarantees the simple analyticity properties of (6) and (7) !).

Unitarity leads to the optical theorem in the following form

$$\mathrm{Im}\{f_{3/2,1/2}(\nu)\} = \frac{\nu}{4\pi} \sigma_{3/2,1/2}(\nu) , \tag{10}$$

where $\sigma_{3/2}$ ($\sigma_{1/2}$) denotes the total cross section for photon nucleon scattering with parallel (antiparallel) spins. From (8a,b) and (10) we find the desired relation

$$\mathrm{Im}\{f_1(\nu)\} = \frac{\nu}{4\pi} \frac{1}{2} [\sigma_{1/2}(\nu) + \sigma_{3/2}(\nu)]$$

$$= \frac{\nu}{4\pi} \sigma_{tot}(\nu) \tag{11a}$$

$$\mathrm{Im}\{f_2(\nu)\} = \frac{\nu}{4\pi} \frac{1}{2} [\sigma_{1/2}(\nu) - \sigma_{3/2}(\nu)] . \tag{11b}$$

From the diffraction picture of high-energy scattering one expects that the total cross section $\sigma_{tot}(\nu)$ will tend to a non-vanishing constant $\sigma_{tot}(\infty)$ for high energies. This was indeed observed with $\sigma_{tot}(\infty) \simeq 100\ \mu b$ [9]; see Fig.1, where the total absorption cross section $\sigma_{tot}(\gamma + p \to \text{hadrons})$ is shown up to 18 GeV.

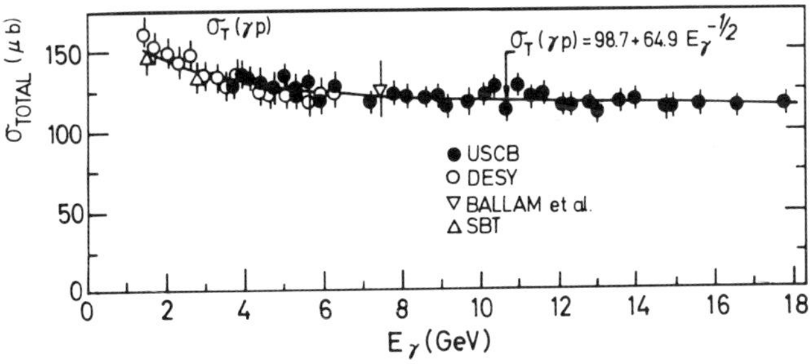


Fig.1 The total photon absorption cross section $\sigma_{tot}(\gamma + p \to \text{hadrons})$ [9]

In view of this fact and (11a) the integral in (6) does not converge. But due to Froissart's bound $\sigma_{tot}(\nu) \leq \text{const.} \log^2 \nu$ the once subtracted version of (6) makes sense in any case.

$$\text{Re}\{f_1(\nu)\} = f_1(0) + \frac{2\nu^2}{\pi} P \int_{\nu_0}^{\infty} \frac{\text{Im}\{f_1(\nu')\}}{\nu'^2 - \nu^2} \frac{d\nu'}{\nu'} .$$

Using (11a) and the low-energy theorem we have the final form of the dispersion relation for $f_1$

$$\text{Re}\{f_1(\nu)\} = -\frac{\alpha}{m} + \frac{\nu^2}{2\pi^2} P \int_{\nu_0}^{\infty} \frac{\sigma_{tot}(\nu')}{\nu'^2 - \nu^2} d\nu' . \tag{12}$$

This relation is known as the Kramers-Kronig relation, since it was already derived by these physicists within classical optics.

Since this relation is based on analyticity properties of $f_1$, which can be exactly proved by axiomatic field theory, and on very well-founded assumptions on the high-energy behavior, an experimental check of (12) is of basic interest. The rhs can be readily calculated [10] using a convenient parametrization of the data given in Fig.1. But only recently the first experimental point for the real part $\text{Re}\{f_1(\nu)\}$ has been obtained at $\nu = 2.2$ GeV by measuring the interference of the Bethe-Heitler $e^+e^-$-production with the semi-virtual Compton scattering for small final photon masses $k'^2$; see Fig.2. The result [11] coincides with the prediction of dispersion theory within the large errors; see Fig.3.

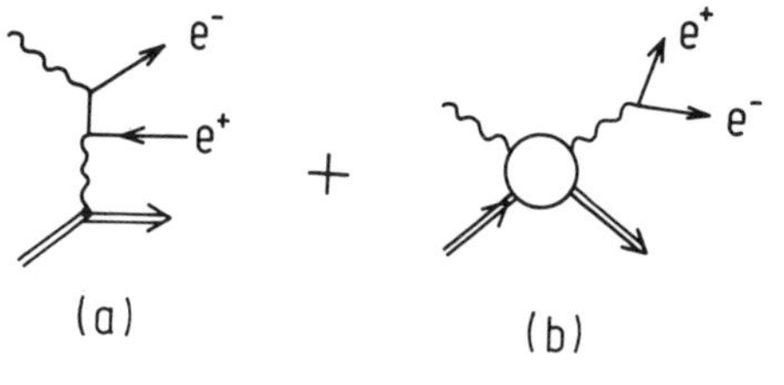


Fig.2 The superposition of the Bethe-Heitler process (a) with semi-virtual Compton scattering (b)

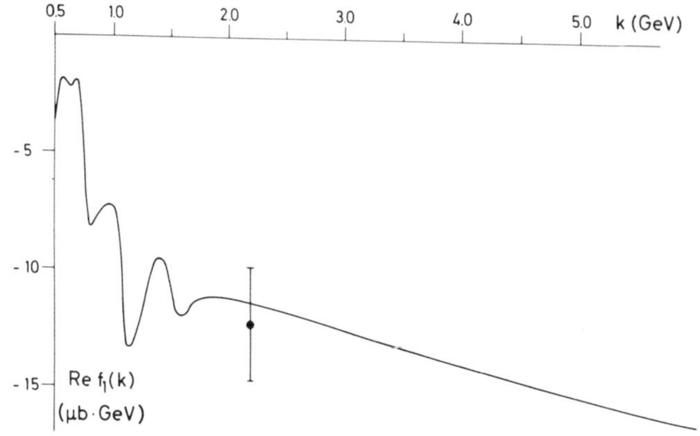


Fig.3 The first test of the Kramers-Kronig relation [11]

Turning now to the dispersion relation (7) for $f_2$, one has good hopes that the integral converges without subtraction. According to folklore the diffraction scattering is spin independent; thus one expects the vanishing of

$$\sigma_{1/2}(\nu) - \sigma_{3/2}(\nu) \underset{\nu\to\infty}{\to} 0 \tag{13}$$

which leads to a convergent expression on the rhs of (7). But putting $\nu = 0$ in this equation and using the low-energy theorem (5a) and the optical theorem (11b) one finds [12,13]

$$\frac{2\pi^2\alpha}{m^2}\kappa^2 = \int_{\nu_0}^{\infty} \frac{\sigma_{3/2}(\nu) - \sigma_{1/2}(\nu)}{\nu}\, d\nu \ . \tag{14}$$

This Drell-Hearn-Gerasimov sum rule has been checked repeatedly [5,12,14] for the proton as target and found to be reasonably satisfied. As long as no measurements for Compton scattering of circularly polarized photons off polarized protons are available the necessary information about the integrand must be gained by summing over all partial cross sections

$$\sigma_{1/2,3/2}(\gamma p \to \text{hadrons}) = \sigma_{1/2,3/2}(\gamma p \to \pi N) + \sigma_{1/2,3/2}(\gamma p \to 2\pi N) + \cdots . \quad (15)$$

But also the cross sections on the rhs have not been measured directly. A multipole analysis of single pion production is necessary to get values for the first term in (15).

As long as the second and higher terms in (15) can be neglected the sum rule contains the qualitative rule: the helicity 3/2 cross section for photopion production weighted with $1/\nu$ must be larger than the helicity 1/2 cross section. This fits very well with the fact that the 2nd ($D_{13}$) and 3rd ($F_{15}$) resonance are excited in the helicity 3/2 state and that also for the 1st ($P_{33}$) resonance $\sigma_{3/2} > \sigma_{1/2}$ is valid.

In the detailed analysis also model calculations for two pion production have been done. Stating the result of a larger story briefly one finds that for the proton the rhs of (14) is by 30% larger (namely 260 $\mu$b) than $(2\pi^2\alpha/m^2)\kappa_p^2 = 205$ $\mu$b [14].

This difficulty becomes more transparent if one decomposes both sides of (14) according to their isospin content[1]

$$\kappa_{p,n}^2 = (\kappa_s \pm \kappa_v)^2 = \kappa_s^2 \pm 2\kappa_s\kappa_v + \kappa_v^2$$

$$\sigma_{1/2,3/2}^{p,n} = \sigma_{1/2,3/2}^{ss} \pm 2\sigma_{1/2,3/2}^{sv} + \sigma_{1/2,3/2}^{vv} . \quad (16)$$

The second terms on the rhs of both equations are due to an interference of isoscalar and isovector photons. As was first observed by FOX and FREEDMAN [5] and now found again [14], there is serious trouble for the S-V interference sum rule. The integral gives +39 $\mu$b, while one expects -14.7 $\mu$b. If this will be confirmed, it may be necessary to change appreciably our picture for high-energy Compton scattering. In the Regge language this failure of the Drell-Hearn sum rule may point to the existence of a fixed J-pole with the quantum numbers of the A1-meson; $J^{PG} = 1^{+-}$, see [5,15]. Further considerations about fixed poles can be found in Sec.2.3.

Going back to the Kramers-Kronig relation we find an explicit expression for the polarizability (5b) by comparing (5) with (12)

---

[1] Compare Sec.1.2.

$$\alpha + \beta = \frac{1}{2\pi^2} \int_{\nu_0}^{\infty} \frac{\sigma_{tot}(\nu)}{\nu^2} d\nu \ . \tag{17}$$

A direct evaluation of the dispersion integral in (17) using experimental data up to 30 GeV and the usual Regge extrapolation for higher energies yields [6]

$$\alpha_p + \beta_p = (14.1 \pm 0.3) \times 10^{-43} \ \mathrm{cm}^3 \ . \tag{17a}$$

The difference between this result and the direct measurement of $\alpha_p + \beta_p$ may be attributed to an additional term in the sum rule (17) originating from a Kronecker delta contribution to the Compton amplitude in the complex angular momentum plane.[2]

The difference $\alpha_p - \beta_p$ may be estimated by means of backward dispersion relations. Saturating the resulting sum rule by means of $\pi N$- and $\pi\Delta$-intermediate states in the direct channel and neglecting the contributions from the $\gamma\gamma \to N\bar{N}$ annihilation channel one obtains the estimate [17]

$$\alpha_p - \beta_p \simeq -6 \times 10^{-43} \ \mathrm{cm}^3 \ , \tag{17b}$$

in complete disagreement with the direct measurement.

In view of the classical result of atomic physics [16]

$$\alpha \simeq (\mathrm{radius})^3 \simeq 10^{-39} \ \mathrm{cm}^3 \ ,$$

the proton must be regarded as an electrically rigid object.

To conclude this survey on the basic aspects of Compton scattering we pay attention to the *current algebra sum rules*. As we shall write down explicitly in (21) the Compton amplitude is closely connected with the commutator of two electromagnetic currents. Therefore assumptions about these commutators as discussed in current algebra often lead to direct or indirect consequences for Compton cross sections. For illustration we consider one sum rule.

We start from the commutator between the electrical dipole moment operator of the isovector part of the current

---

[2] Compare [5] and the literature cited therein.

$$D_i = \int d^3x \, x \, J_i^{0,V}(x)$$

where i = 1,2,3 label the isospin components. Current algebra [18] leads to

$$[D_1 + iD_2, D_1 - iD_2] = 2 \int d^3x \, x^2 \, J_3^{0,V}(x) \tag{18}$$

where the commutator is taken for equal times and the dot-product between the two spatial vectors $D_1 \pm iD_2$ is understood. By taking nucleon matrix elements from (18) one finds after some manipulations the Cabbibo-Radicati sum rule [18]

$$\frac{1}{3} \langle r^2 \rangle^V - \left(\frac{\kappa^V}{2m}\right)^2 = \frac{1}{2\pi^2 \alpha} \int_{\nu_0}^{\infty} \frac{d\nu}{\nu} \left[2\sigma^V_{I=1/2}(\nu) - \sigma^V_{I=3/2}(\nu)\right] \; . \tag{19}$$

On the lhs the isovector anomalous magnetic moment ($\kappa^V = 3.7$) and the isovector charge radius

$$\langle r^2 \rangle^V \underset{\text{def.}}{=} 6F_1^{V\prime}(0)$$

enter ($F_1^V(q^2)$ = isovector Dirac form factor of the nucleon). The integrand contains the total cross section for isovector photons hitting nucleons and leading to total isospin states I = 1/2 or I = 3/2, respectively. A numerical estimate of the integral has been given in [5] and quite recently by GENZ [19]. In both references the authors obtain rough agreement with the experimental value of the lhs of (19) [5,19].

There exist quite a number of other Compton scattering sum rules. The J = 0 fixed pole sum rule will be discussed in Section 2.3. For a systematic treatment of all other sum rules compare [5].

## 1.2 Kinematics

In this section we collect the kinematical definitions and relations necessary for a complete description of Compton scattering on nucleons. Using the notation of Fig.4 we write for the relevant S-matrix element

$$\langle k',p'|S|k,p\rangle = \langle k'|k\rangle\langle p'|p\rangle - i(2\pi)^4 \delta^{(4)}(k+p-k'-p')T(k',p',k,p) \tag{20}$$

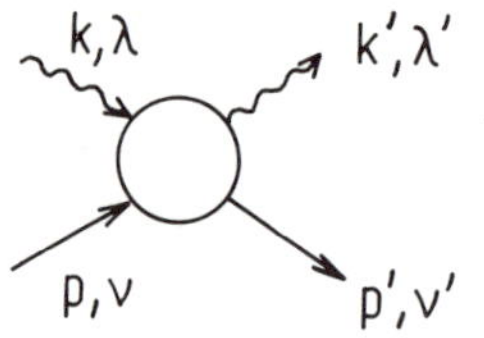


Fig.4 Kinematical notation for Compton scattering (k,k'; p,p' are momenta; λ,λ'; ν,ν' helicities)

where we have for the moment suppressed helicity indices. Here invariant normalization

$$\langle p'|p\rangle = (2\pi)^3 2p_0 \delta^{(3)}(\underset{\sim}{p} - \underset{\sim}{p}')$$

has been used.

We introduce the usual Chew variables $s = W^2 = (p+k)^2$, $t = (k-k')^2$, $u = (p-k')^2$ and $\nu = (1/4m)(s-u)$ (our metric is $g_{oo} = -g_{ii} = 1$). Note that $\nu$ reduces for forward scattering to that expression given in Section 1.1.

By the usual reduction formulae the Compton amplitude can be expressed by a retarded commutator of electromagnetic current operators $J_\mu$ [20,21]

$$T(k',\lambda',p',\nu';k,\lambda,p,\nu)$$

$$= -i4\pi\alpha\varepsilon^{\mu *}_{(\lambda')}\varepsilon^{\nu}_{(\lambda)} \int d^4x \langle p',\nu'|\Theta(x_0)[J_\mu(x),J_\nu(0)]|p,\nu\rangle \; . \tag{21}$$

Here $\varepsilon^{\nu}_{(\lambda)}$ denotes the polarization four vector for a photon with helicity $\lambda$.

The *isotopic spin decomposition* can be most simply written as

$$T = T_0 + T_1\tau_3 \tag{22}$$

so that

$$\langle\gamma p|T|\gamma p\rangle = T_0 + T_1 \; , \qquad \langle\gamma n|T|\gamma n\rangle = T_0 - T_1 \; . \tag{22a,b}$$

On the other hand by $J_\mu = J^S_\mu + j^V_\mu$ one finds

$$T = T^{SS} + T^{VV} + T^{SV} \tag{23}$$

so that $T_0$ is given by the sum of the isoscalar-isoscalar and the isovector-isovector term and $T_1$ by the isoscalar-isovector interference.[3] Formula (23) has already been used in (16) of the first section.

To discuss the spin content of T we use helicity amplitudes [22] in the CMS by choosing appropriate polarization vectors $\varepsilon^\mu$ and nucleon states. We denote the (s-channel) helicity amplitudes by

$$M_{\lambda',\nu';\lambda,\nu} \quad \text{with} \quad \begin{cases} \lambda,\lambda' = \pm 1 \\ \nu,\nu' = \pm 1/2 \end{cases}$$

which are normalized in accordance with JACOB and WICK [22], i.e., $M = -(8\pi W)^{-1} T_{CMS}$.

These 16 amplitudes depend on the Lorentz-invariant quantities s and t. By parity invariance

$$M_{-\lambda',-\nu';-\lambda,-\nu} = (-1)^{\lambda-\nu-\lambda'+\nu'} M_{\lambda',\nu';\lambda,\nu}$$

only 8 amplitudes are independent. Time-reversal invariance

$$M_{\lambda,\nu;\lambda',\nu'} = (-1)^{\lambda-\nu-\lambda'+\nu'} M_{\lambda',\nu';\lambda,\nu}$$

gives another 2 constraints for this elastic process. Thus we are left with 6 different helicity amplitudes for which we take

$$\begin{aligned} &\Phi_1 = M_{1\frac12,1\frac12}\,, \quad \Phi_2 = M_{-1-\frac12,1\frac12}\,, \quad \Phi_3 = M_{-1\frac12,1\frac12}\,, \\ &\Phi_4 = M_{1-\frac12,1\frac12}\,, \quad \Phi_5 = M_{1-\frac12,1-\frac12}\,, \quad \Phi_6 = M_{-1\frac12,1-\frac12}\,. \end{aligned} \tag{24}$$

By angular momentum conservation in the forward direction

$$\Theta = 0 \quad \text{only} \quad \Phi_1 \quad \text{and} \quad \Phi_5 \tag{24a}$$

in the backward direction

$$\Theta = \pi \quad \text{only} \quad \Phi_2 \quad \text{and} \quad \Phi_6 \tag{24b}$$

[3] In $T^{VV}$ no total isospin $I = 1$ is contained since $J_\mu$ carries only $I_3 = 0$ components.

contribute. There is no hope that the $2\times 6-1=11$ independent experiments necessary for a complete empirical determination of the $\Phi_i$ will ever be carried out. The formulae for the three classical observables are:

Differential cross section

$$\frac{d\sigma}{d\Omega}=\frac{1}{2}\left(|\Phi_1|^2+|\Phi_2|^2+2|\Phi_3|^2+2|\Phi_4|^2+|\Phi_5|^2+|\Phi_6|^2\right) \quad . \tag{25}$$

The additional factors 2 arise because these amplitudes occur twice because of time-reversal invariance.

Photon asymmetry

$$\frac{d\sigma}{d\Omega}\Sigma = \frac{1}{2}\left(\frac{d\sigma_\perp}{d\Omega}-\frac{d\sigma_\parallel}{d\Omega}\right) = \mathrm{Re}\{(\Phi_1+\Phi_5)^*\Phi_3+(\Phi_2-\Phi_6)^*\Phi_4\}.. \tag{26}$$

Polarization of the recoiling nucleon

$$\frac{d\sigma}{d\Omega}P = \mathrm{Im}\{(\Phi_1+\Phi_5)^*\Phi_4-(\Phi_2-\Phi_6)^*\Phi_3\} \quad . \tag{27}$$

Because of time-reversal invariance the target polarization T is equal to P.

Partial wave expansions can be found by the Jacob-Wick formula

$$M_{\lambda',\nu';\lambda,\nu} = \sum_J (2J+1)M^J_{\lambda',\nu';\lambda,\nu}d^J_{\lambda-\nu,\lambda'-\nu'}(\Theta,\phi) \quad .$$

As in photoproduction linear combinations of the $M^J_{\lambda',\nu';\lambda,\nu}$ are used, namely multipole amplitudes

$$f^{L\pm}_{EE}\ , \quad f^{L\pm}_{MM}\ , \quad f^{L\pm}_{EM}\ . \tag{28}$$

Here L is the total photon angular momentum and the physical meaning of these amplitudes is as follows: $f^{L\pm}_{EM}$ describes the transition between an electrical $2^L$-pole in the initial state to a magnetic $2^{L'}$-pole of the same parity ($L'=L\pm 1$) in the final state. The total angular momentum of the $\gamma$-N system is given by $J=L\pm 1/2$, wherefrom also L' is uniquely determined.

For the two other types of amplitudes an analogous interpretation can be given. To illustrate this definition we remark that time reversal leads to the equations

$$f_{EM}^{L+} = f_{ME}^{(L+1)-} \quad \text{and} \quad f_{EM}^{L-} = f_{ME}^{(L-1)+} \,. \tag{29}$$

Their physical content is simply that one can change the initial and the final state without changing the amplitude.

As an example for the multipole expansion we write down the formula for $\Phi_1$:

$$\Phi_1 = \frac{1}{2}\sum_{L=0}^{\infty} (L+1) \Big\{ (L+2)^2 \left( f_{EE}^{(L+1)-} + f_{MM}^{(L+1)-} \right) + L^2 \left( f_{EE}^{L+} + f_{MM}^{L+} \right)$$

$$- 2L(L+2) \left( f_{EM}^{L+} + f_{ME}^{L+} \right) \Big\} d_{1/2,1/2}^{L+1/2}(\Theta,\phi) \tag{30}$$

with $f_{EE}^{0+} = f_{MM}^{0+} = f_{ME}^{0+} = f_{EM}^{0+} \underset{\text{def.}}{=} 0$. The complete set of expansions is given in Table 1. Special cases will be treated in Section 1.6.

Table 1 Multipole expansion of the helicity amplitudes

$$\Phi_{\substack{1\\2}} = \frac{1}{2}\sum_{L=0}^{\infty} (L+1) \Big\{ (L+2)^2 \left( f_{EE}^{(L+1)-} \pm f_{MM}^{(L+1)-} \right) \pm L^2 \left( f_{EE}^{L+} \pm f_{MM}^{L+} \right)$$

$$\mp 2L(L+2) \left( f_{EM}^{L+} \pm f_{ME}^{L+} \right) \Big\} d_{1/2,\pm 1/2}^{L+1/2}$$

$$\Phi_{\substack{3\\4}} = \frac{1}{2}\sum_{L=1}^{\infty} (L+1)\sqrt{L(L+2)} \Big\{ (L+2) \left( f_{EE}^{(L+1)-} \mp f_{MM}^{(L+1)-} \right) \pm L \left( f_{EE}^{L+} \mp f_{MM}^{L+} \right)$$

$$\mp 2 \left( f_{EM}^{L+} \mp f_{ME}^{L+} \right) \Big\} d_{1/2,\mp 3/2}^{L+1/2}$$

$$\Phi_{\substack{5\\6}} = \frac{1}{2}\sum_{L=1}^{\infty} (L+1)L(L+2) \Big\{ \left( f_{EE}^{(L+1)-} \pm f_{MM}^{(L+1)-} \right) \pm \left( f_{EE}^{L+} \pm f_{MM}^{L+} \right)$$

$$\pm 2 \left( f_{EM}^{L+} \pm f_{ME}^{L+} \right) \Big\} d_{3/2,\pm 3/2}^{L+1/2}$$

with $f_{EE}^{0+} = f_{MM}^{0+} = f_{EM}^{0+} = f_{ME}^{0+} \underset{\text{def.}}{=} 0$

## 1.3 Invariant Amplitudes[4]

For the treatment of analyticity it is more convenient to work with a set of invariant amplitudes free of kinematic singularities and zeros. For their definition we write the Compton amplitude T in the form

$$T = \varepsilon^{\mu *}_{(\lambda')} \varepsilon^{\nu}_{(\lambda)} \bar{u}_{(\nu')}(p') T_{\mu\nu} u_{(\nu)}(p) \tag{31}$$

with the normalization of the Dirac spinors $\bar{u}u = 2m$. The tensor $T_{\mu\nu}$ may be expanded with respect to a tensor basis $I^i_{\mu\nu}$

$$T_{\mu\nu} = \sum_{i=1}^{6} A_i I^i_{\mu\nu} \tag{32}$$

where the invariant amplitudes $A_i$ depend on the Lorentz invariant variables s, t and u with the constraint $s + t + u = 2m^2$. Here 6 terms occur corresponding to the 6 independent helicity amplitudes. For an explicit construction of $I^i_{\mu\nu}$ we introduce the four vectors

$$P_\mu = \frac{1}{2}(p_\mu + p'_\mu) \ , \quad K_\mu = \frac{1}{2}(k_\mu + k'_\mu) \ , \quad Q_\mu = k_\mu - k'_\mu = p'_\mu - p_\mu \ .$$

In order to avoid problems due to subsidiary conditions connected with gauge invariance, we first consider the scattering of non-zero mass vector particles off nucleons. The corresponding set of 12 independent covariants $\hat{I}^i_{\mu\nu}$ are supposed to contain a minimal number of momenta (minimal polynomial expansion of $T_{\mu\nu}$) and to be time-reversal invariant.

Let us write in this case

$$T_{\mu\nu} = \sum_{i=1}^{12} C_i \hat{I}^i_{\mu\nu} \tag{33}$$

with [23]

$$\begin{aligned}
&\hat{I}^1_{\mu\nu} = g_{\mu\nu} \ , \quad \hat{I}^2_{\mu\nu} = P_\mu P_\nu , \quad \hat{I}^3_{\mu\nu} = P_\mu \gamma_\nu + \gamma_\mu P_\nu \ , \quad \hat{I}^4_{\mu\nu} = P_\mu P_\nu \not{K} \ ,\\
&\hat{I}^5_{\mu\nu} = \gamma_\mu \gamma_\nu - \gamma_\nu \gamma_\mu \ , \quad \hat{I}^6_{\mu\nu} = \gamma_\mu \not{K} \gamma_\nu - \gamma_\nu \not{K} \gamma_\mu \ , \quad \hat{I}^7_{\mu\nu} = g_{\mu\nu} \not{K} \ ,\\
&\hat{I}^8_{\mu\nu} = P_\mu \not{K} \gamma_\nu + \gamma_\mu \not{K} P_\nu \ , \quad \hat{I}^9_{\mu\nu} = k_\mu k'_\nu \ , \quad \hat{I}^{10}_{\mu\nu} = P_\mu k'_\nu + k_\mu P_\nu \ ,\\
&\hat{I}^{11}_{\mu\nu} = k_\mu \gamma_\nu + \gamma_\mu k'_\nu \ , \quad \hat{I}^{12}_{\mu\nu} = \gamma_\nu \not{K} k_\mu + k'_\nu \not{K} \gamma_\mu \ .
\end{aligned} \tag{34}$$

[4] The reader who is not interested in the technical details for deriving invariant amplitudes free of kinematical singularities, etc., may skip this section.

It is obvious that our $\hat{I}^i_{\mu\nu}$ satisfy the time-reversal invariance condition

$$\hat{I}^i_{\mu\nu}(k',p';k,p) = \mathbb{K}\,\hat{I}^{iT}_{\nu\mu}(k,p;k',p')\,\mathbb{K}^{-1}$$

where the Dirac-matrix $\mathbb{K}$ is defined by the relation[5]

$$\mathbb{K}\,\gamma^T_\mu\,\mathbb{K}^{-1} = \gamma_\mu \ .$$

Going now back to Compton scattering we observe that the amplitude (32) must, due to current conservation, satisfy the gauge invariance conditions

$$k'_\mu T^{\mu\nu} = 0 = T^{\mu\nu}k_\nu \ . \tag{35}$$

Eq.(35) is satisfied if and only if $T_{\mu\nu}$ is invariant with respect to the application of the gauge projection operator

$$G_{\mu\nu} := g_{\mu\nu} - \frac{k_\mu k'_\nu}{k\cdot k'} \ ,$$

i.e.,

$$(GTG)_{\mu\nu} = T_{\mu\nu} \ .$$

Therefore, the most convenient way of imposing gauge invariance on our scattering amplitude is by operating on the $\hat{I}^i_{\mu\nu}$ with the gauge projection operator G [25,26]

$$\tilde{I}^i_{\mu\nu} := (G\hat{I}^iG)_{\mu\nu} \ . \tag{36}$$

As the gauge projections of $k_\mu$ and $k'_\nu$ vanish

$$G_{\mu\nu}k^\nu = k'^\mu G_{\mu\nu} = 0$$

we obtain

$$\tilde{I}^i_{\mu\nu} = 0 \quad \text{for } i = 9,\cdots,12.$$

The resulting set of eight gauge invariant $\tilde{I}^i_{\mu\nu}$ is not yet satisfying, because (in the following we restrict ourselves to real Compton scattering, i.e., $k^2 = k'^2 = 0$)

i) we need only 6 independent amplitudes,

ii) the $\tilde{I}^i_{\mu\nu}$ contain kinematic singularities (single and double poles) at $t = 0$, leading to the occurrence of kinematic zeros in the corresponding invariant amplitudes.

---

[5] For the usual representation of the $\gamma_\mu$ as given for example in BJORKEN-DRELL's textbook [24] $\mathbb{K}$ is given by $i\gamma_1\gamma_3$ (called T in [24]).

An alternative construction of a set of six independent, time-reversal and gauge invariant covariants has been given by PRANGE [27] and by HEARN and LEADER [28], respectively.

Starting from the time-like vector $K_\mu$ ($K^2 > 0$) one looks for four mutually orthogonal vectors. A possible set is given by

$$P'_\mu = P_\mu - \frac{P\cdot K}{K^2} K_\mu \;, \quad K_\mu \;, \quad Q_\mu \;, \quad \text{and} \quad N_\mu = \varepsilon_{\mu\nu\rho\sigma} P'^\nu K^\rho Q^\sigma \;.$$

P', Q and N are space-like vectors because of their orthogonality with K. With the help of these vectors we define according to [27,28]

$$\check{I}^1_{\mu\nu} = \frac{P'_\mu P'_\nu}{-P'^2} \;, \quad \check{I}^4_{\mu\nu} = \frac{P'_\mu P'_\nu}{-P'^2} \not{K} \;, \quad \check{I}^2_{\mu\nu} = \frac{N_\mu N_\nu}{-N^2} \;, \quad \check{I}^5_{\mu\nu} = \frac{N_\mu N_\nu}{-N^2} \not{K} \;,$$

$$\check{I}^3_{\mu\nu} = \gamma_5 \frac{P'_\mu N_\nu - P'_\nu N_\mu}{\sqrt{P'^2 N^2}} \;, \quad \check{I}^6_{\mu\nu} = \check{I}^3_{\mu\nu} \not{K} \;.$$

Gauge invariance of the $\check{I}^i_{\mu\nu}$ is guaranteed by this choice. In fact by definition of K and Q one has

$$k_\mu = K_\mu + Q_\mu \;, \quad k'_\mu = K_\mu - Q_\mu$$

and (35) holds for each $\check{I}^i_{\mu\nu}$ as a direct consequence of the orthogonality properties of our four basis vectors.

Compared with the $\tilde{I}^i_{\mu\nu}$ these covariants have the disadvantage that—due to the $P'^2$ and $N^2$ in the denominators—kinematic singularities not only occur at $t = 0$, but also at $su = m^4$. Problem i) is easily solved by means of the orthogonal tensor basis $\tilde{I}^i_{\mu\nu}$ given above.

One derives the following two identities:

$$mt\tilde{I}^5_{\mu\nu} = 4\nu m\tilde{I}^6_{\mu\nu} , \quad 2m\tilde{I}^8_{\mu\nu} = m\nu\tilde{I}^5_{\mu\nu} + P^2\tilde{I}^6_{\mu\nu} \;. \tag{37a,b}$$

From (37b) we can eliminate $\tilde{I}_{\mu\nu}$ without introducing kinematic singularities.

On a first sight one might think that problem ii) can be solved by multiplying each covariant $\tilde{I}^i_{\mu\nu}$ by the appropriate factor of t or $t^2$. But this would be an overcompensation leading to kinematic singularities in the corresponding invariant functions, because there exist linear combinations of the $\tilde{I}^i_{\mu\nu}$ which are less singular. Therefore, our next task is to find these appropriate linear combinations.

The covariants $\tilde{I}^{1,6,7}_{\mu\nu}$ do not contain any $t^{-2}$-term. By means of the following combination we may kill the $t^{-2}$-terms in the remaining covariants [26].

$$\tilde{I}^i_{\mu\nu} \rightarrow \tilde{I}^i_{\mu\nu} + \frac{k' \cdot \hat{I}^i \cdot k}{k \cdot k'} \tilde{I}^1_{\mu\nu}$$

$$= \hat{I}^i_{\mu\nu} - \frac{1}{k \cdot k'} \left\{ k_\mu \left( k'^\lambda \hat{I}^i_{\lambda\nu} \right) + \left( \hat{I}^i_{\mu\lambda} k^\lambda \right) k'_\nu - (k' \cdot \hat{I}^i \cdot k) g_{\mu\nu} \right\} . \quad (38)$$

Now we have to express the terms $(k' \cdot \hat{I}^i \cdot k)\tilde{I}^1_{\mu\nu}$ on the lhs of (38) as linear combinations of $\tilde{I}^1_{\mu\nu}$. By means of (34) and (36) we obtain

$$k' \cdot \hat{I}^i \cdot k\tilde{I}^1_{\mu\nu} = \begin{cases} m^2\nu^2\tilde{I}^1_{\mu\nu} & \text{for } i=2 \\ 2\nu m\tilde{I}^7_{\mu\nu} & \text{for } i=3 \\ m^2\nu^2\tilde{I}^7_{\mu\nu} & \text{for } i=4 \\ -\frac{1}{4} t\tilde{I}^1_{\mu\nu} - 4\left(m\tilde{I}^7_{\mu\nu} - m\nu\tilde{I}_{\mu\nu}\right) & \text{for } i=5 . \end{cases} \quad (39)$$

In evaluating the last line we used the fact that the $\tilde{I}^i_{\mu\nu}$ are sandwiched between Dirac spinors. For $\tilde{I}^4_{\mu\nu}$, and only in that case, we may even find a linear combination which is free of both singularities, the linear and the quadratic ones [26].

$$\tilde{I}^4_{\mu\nu} \rightarrow 4\tilde{I}^4_{\mu\nu} - m\nu\left(2\tilde{I}^3_{\mu\nu} - \tilde{I}^6_{\mu\nu}\right) . \quad (40)$$

The remaining linear singularities at $t=0$ in the combinations (38) for $i=$ 2,3,5 and in $\tilde{I}^{1,6,7}_{\mu\nu}$ can be killed only by multiplying them by the factor t.

Finally we arrive by means of the procedure described above at the following minimal polynomial set of seven gauge-invariant covariants $I^i_{\mu\nu}$ free of kinematic singularities [23]:

$$I^1_{\mu\nu} = k \cdot k'\tilde{I}^1_{\mu\nu} = g_{\mu\nu} k \cdot k' - k^\mu k'^\nu ,$$

$$I^2_{\mu\nu} = k \cdot k'\tilde{I}^2_{\mu\nu} + m^2\nu^2\tilde{I}^1_{\mu\nu} = k \cdot k'P_\mu P_\nu - m\nu(k_\mu P_\nu + P_\mu k'_\nu) + \nu^2 g_{\mu\nu} ,$$

$$I^3_{\mu\nu} = k \cdot k'\tilde{I}^3_{\mu\nu} + 2m\nu\tilde{I}^7_{\mu\nu}$$

$$= k \cdot k'(P_\mu\gamma_\nu + \gamma_\mu P_\nu) - m\nu(k_\mu\gamma_\nu + k'_\nu\gamma_\mu) - (k_\mu P_\nu + P_\mu k'^\nu - 2m\nu g_{\mu\nu})K\!\!\!/ ,$$

$$I^4_{\mu\nu} = 4\tilde{I}^4_{\mu\nu} - m\nu\left(2\tilde{I}^3_{\mu\nu} - \tilde{I}^6_{\mu\nu}\right) - k \cdot k'\tilde{I}^7_{\mu\nu}$$

$$= 4K\!\!\!/ P_\mu P_\nu - m\nu[2(P_\mu\gamma_\nu + \gamma_\mu P_\nu) - \gamma_\mu K\!\!\!/ \gamma_\nu + \gamma_\nu K\!\!\!/ \gamma_\mu] , \quad (41)$$

$$I^5_{\mu\nu} = -k \cdot k'\tilde{I}^5_{\mu\nu} + 4\left(m\tilde{I}^7_{\mu\nu} - m\nu\tilde{I}^1_{\mu\nu}\right)$$

$$= -k \cdot k'[\gamma_\mu,\gamma_\nu] + k_\mu[K\!\!\!/,\gamma_\nu] + [\gamma_\mu,K\!\!\!/]k'_\nu + 4g_{\mu\nu}(mK\!\!\!/ - \nu m) \quad , \tag{41}$$

$$I^6_{\mu\nu} = -k \cdot k'\tilde{I}^6_{\mu\nu}$$

$$= k \cdot k'(\gamma_\nu K\!\!\!/\gamma_\mu - \gamma_\mu K\!\!\!/\gamma_\nu) + 2m\nu(k_\mu\gamma_\nu + \gamma_\mu k'_\nu) - 2K\!\!\!/(k_\mu P_\nu + P_\mu k'_\nu) \quad ,$$

$$\tilde{I}^7_{\mu\nu} = t\tilde{I}^7_{\mu\nu} \quad .$$

The identity (37a) expressed in terms of the $I^i_{\mu\nu}$ looks as follows:

$$2m^2 I^7_{\mu\nu} = -4m^2\nu I^1_{\mu\nu} + \frac{1}{2}mtI^5_{\mu\nu} - 2m\nu I^6_{\mu\nu} \quad . \tag{42}$$

This relation allows us to eliminate $I^7_{\mu\nu}$ without introducing kinematic singularities.

According to the construction of the corresponding tensor basis $I^i_{\mu\nu}$ the invariant amplitudes $A_i$ $(i = 1 \cdots 6)$ are free of kinematic singularities and zeros.

As different authors use different linear combinations for the final form of the $I^i_{\mu\nu}$, we list in the following table the relations between the corresponding invariant amplitudes.

Table 2 Relations between invariant amplitudes for different choices of tensor bases in (33)

| $A_i$ - KELLET [23] | $A'_i$ - JONES-SCADRON [25] | $\bar{A}_i$ - BARDEEN-TUNG [26] |
|---|---|---|
| $A_1$ | $2A'_1 + \frac{1}{4}(t - 4m^2)A'_2 + \nu A'_4$ | $\frac{\bar{A}_1}{2} - \frac{1}{4}\left(m^2 - \frac{t}{4}\right)\bar{A}_5 - \frac{m}{2}\bar{A}_4 - \frac{\nu}{4}\bar{A}_6$ |
| $A_2$ | $2A'_2$ | $\bar{A}_5/2$ |
| $A_3$ | $2A'_3$ | $\bar{A}_4/2$ |
| $A_4$ | $-A'_4$ | $\bar{A}_6/4$ |
| $A_5$ | $-2A'_5 - \frac{t}{8m}A'_4$ | $\bar{A}_3/4 + \frac{\bar{A}_6}{8m}\left(\frac{t}{4} - m^2\right)$ |
| $A_6$ | $-2A'_6 + \frac{\nu}{2m}A'_4$ | $-\frac{\bar{A}_2}{4} - \frac{\nu}{8m}\bar{A}_6$ |

The relations between the invariant amplitudes of BARDEEN-TUNG and those of HEARN-LEADER are given in [26].

By means of an appropriate choice of polarization vectors $\varepsilon$ and Dirac spinors u in (31) and our decomposition (32) it is a straightforward but somewhat tedious task to express helicity amplitudes in terms of invariant amplitudes. Therefore, we will give only the final results.

s-channel helicity amplitudes [23][6]

$$f^s_{1\frac{1}{2},1\frac{1}{2}} = -2k^2(\cos\tfrac{1}{2}\Theta)\left[ms(\cos^2\tfrac{1}{2}\Theta)A_2 + 2[s - m^2(\sin^2\tfrac{1}{2}\Theta)]A_3 - 2[s + m^2(\sin^2\tfrac{1}{2}\Theta)]A_4\right]$$

$$f^s_{1-\frac{1}{2},1-\frac{1}{2}} = -2k^2 s(\cos^3\tfrac{1}{2}\Theta)[mA_2 + 2A_3 + 2A_4] \; ,$$

$$f^s_{1\frac{1}{2},1-\frac{1}{2}} = 2k^2W(\sin\tfrac{1}{2}\Theta)(\cos^2\tfrac{1}{2}\Theta)[EWA_2 + 2m(A_3 + A_4)] \; ,$$

$$f^s_{1\frac{1}{2},-1\frac{1}{2}} = -4k^2(\sin^2\tfrac{1}{2}\Theta)(\cos\tfrac{1}{2}\Theta)[m(A_1 + P^2A_2 + mA_3 + 2A_5) + EWA_4] \; , \qquad (43)$$

$$f^s_{1\frac{1}{2},-1-\frac{1}{2}} = 4k^2(\sin\tfrac{1}{2}\Theta)[E(\sin^2\tfrac{1}{2}\Theta)(A_1 + P^2A_2 + mA_3)$$

$$- W(2 - (\sin^2\tfrac{1}{2}\Theta))(mA_4 + 2A_5) + mk(\sin^2\tfrac{1}{2}\Theta)(A_4 + 2A_6)] \; ,$$

$$f^s_{1-\frac{1}{2},-1\frac{1}{2}} = -4k^2(\sin^3\tfrac{1}{2}\Theta)[E(A_1 + P^2A_2 + mA_3) + m(EA_4 - 2kA_6) - 2(k - E)A_5] \; ,$$

with $P^2 = m^2 - t/4$ according to the definition of $P_\mu$.

As usual E and k are the s-channel cm nucleon energy and momentum, respectively, i.e.,

$$E = \frac{s + m^2}{2W} \; , \qquad k^2 = \frac{(s - m^2)^2}{4s} \; .$$

If we insert this result into (25) we can obtain an expression for $d\sigma/d\Omega$ in terms of the $A_i$. We shall not burden the reader with the explicit general formula. Special cases will be given later.

For completeness we note the relation between these $f^s$ and the forward scattering amplitudes introduced in Section 1.1

$$f_{1/2,3/2}(\nu) = -\frac{1}{8\pi m} f^s_{1\pm\frac{1}{2},1\pm\frac{1}{2}}\Big|_{t=0} \; . \qquad (44)$$

[6] Compared to $T^{CMS}$ and M they are normalized as follows:

$$f^s = T^{CMS} = -8\pi WM \; .$$

t-channel helicity amplitudes [25]

$$f^t_{1,-1;\frac{1}{2},\frac{1}{2}} = -tp\left(\sin^2\frac{\Theta_t}{2}\right)\left(-\frac{p^2}{2}A_2 + mA_3\right) ,$$

$$f^t_{1,-1;-\frac{1}{2},\frac{1}{2}} = 2tp(\sin\Theta_t)\left(\sin^2\frac{\Theta_t}{2}\right)(-qA_3 + pA_4) ,$$

$$f^t_{1,-1;\frac{1}{2},-\frac{1}{2}} = 2tp(\sin\Theta_t)\left(\cos^2\frac{\Theta_t}{2}\right)(-qA_3 - pA_4) ,$$

$$\begin{aligned} f^t_{1,1;-\frac{1}{2},-\frac{1}{2}} = -t\Big[p\left(-A_1 + \frac{p^2}{2}A_2 - mA_3\right) \\ - (p+q)(\cos\Theta_t)\left(\frac{p(q-p)}{m}A_4 + 2A_5\right) - 2mqA_6\Big] , \end{aligned} \tag{45}$$

$$\begin{aligned} f^t_{1,1;\frac{1}{2},\frac{1}{2}} = -t\Big[p\left(-A_1 + \frac{p^2}{2}A_2 - mA_3\right) \\ + (p-q)(\cos\Theta_t)\left(-\frac{p(p+q)}{m}A_4 + 2A_5\right) + 2mqA_6\Big] , \end{aligned}$$

$$f^t_{1,1;-\frac{1}{2},\frac{1}{2}} = mt(\sin\Theta_t)(mA_4 + 2A_5) ,$$

where we denote q and p the t-channel cm momenta of the photon and nucleon, respectively, i.e., $t = 4q^2 = 4(p^2 + m^2)$. $\Theta_t$ is the corresponding scattering angle, i.e., $\cos\Theta_t = m\nu/pq$.

## 1.4 Crossing and Analyticity

The (s - u) *crossing operation* is defined by (see Fig.5)

$$k \leftrightarrow -k' \; ; \quad \varepsilon_{(\lambda)} \leftrightarrow \varepsilon_{(\lambda')} \; ; \quad s \leftrightarrow u \; , \text{ i.e., } \nu \leftrightarrow -\nu \; ; \quad t \leftrightarrow t \; . \tag{46}$$

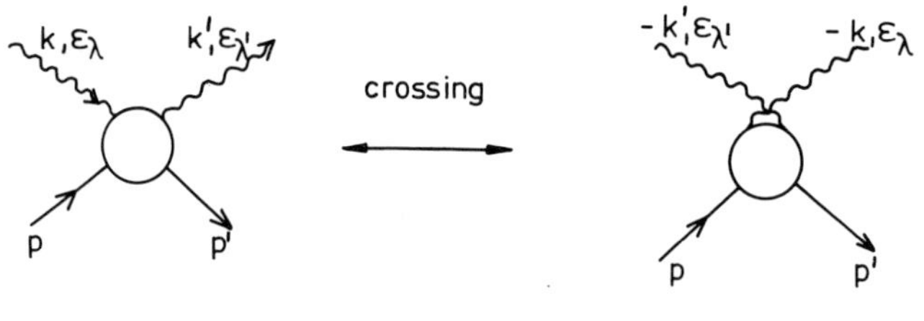


Fig.5 Definition of s - u crossing

From (41) one finds

$$\varepsilon^{*\mu}_{(\lambda')}\varepsilon^{\nu}_{(\lambda)} I^i_{\mu\nu} \overset{\text{crossing}}{\rightarrow} \eta^i_c \varepsilon^{*\mu}_{(\lambda)}\varepsilon^{\nu}_{(\lambda')} I^i_{\mu\nu} \tag{47}$$

with

$$\eta_c^i = \begin{cases} +1 \text{ for } i = 1,2,3,6 \\ -1 \text{ for } i = 4,5 \ . \end{cases}$$

The Compton amplitude does not change under s - u crossing. Thus we have from (32)

$$A_i(u,t,s) = \eta_c^i A_i(s,t,u) \ . \tag{48}$$

The crossing properties (4) of the forward scattering amplitudes $f_1$ and $f_2$ are special cases of (48).

*Analyticity properties* of the amplitudes $A_i$ can be derived rigorously from axiomatic field theory [29]. For fixed t in the interval

$$0 \geq t \geq t_0 \simeq -12.07 \ m_\pi^2$$

the $A_i(s,t)$ are holomorphic functions in the s-plane up to a cut along the real axis. Disregarding subtractions one can summarize these properties by the following dispersion relations

$$A_i(s,t) = A_i^{Born} + A_i^{\eta,\pi^0} + \frac{1}{\pi}\int_{(m+m_\pi)^2}^{\infty} ds' \left[\frac{1}{s'-s} + \eta_c^i \frac{1}{s'-u}\right] \mathrm{Im}\{A_i(s',t)\} . \tag{49}$$

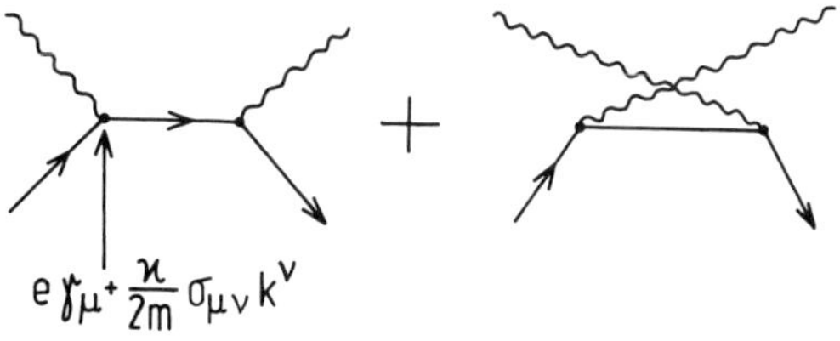


Fig.6 Born diagrams for Compton scattering

The Born terms can be obtained straightforwardly by evaluating the Feynman diagrams of Fig.6. They are given by the following expressions:

$$A_i^{Born} = R_i^+(t)\left[\frac{1}{m^2-s} + \frac{1}{m^2-u}\right] + R_i^-(t)\left[\frac{1}{m^2-s} - \frac{1}{m^2-u}\right]$$

$$+ R_i^{su}\frac{1}{(m^2-s)(m^2-u)} + C_i \tag{50}$$

where the pole residual $R_i$ and the constants $C_i$ are listed in Table 3.

Table 3 Born-term coefficients in (50)

| $A_i$ | $R_i^+(t)$ | $R_i^-(t)$ | $R_i^{su}$ | $C_i$ |
|---|---|---|---|---|
| $A_1$ | $-4\pi\alpha\left[\frac{\kappa}{m}+\frac{\kappa^2}{2m}\left(1-\frac{t}{8m^2}\right)\right]$ | 0 | 0 | $-\frac{\pi\alpha\kappa^2}{m^3}$ |
| $A_2$ | 0 | 0 | $-\frac{16\pi\alpha\kappa}{m}$ | 0 |
| $A_3$ | $\frac{\pi\alpha\kappa^2}{m^2}$ | 0 | $8\pi\alpha(1+\kappa)$ | 0 |
| $A_4$ | 0 | $-\frac{\pi\alpha\kappa^2}{m^2}$ | 0 | 0 |
| $A_5$ | 0 | $\pi\alpha\left[\frac{\kappa}{m}+\frac{\kappa^2}{m}\left(1-\frac{t}{8m^2}\right)\right]$ | 0 | 0 |
| $A_6$ | $-\pi\alpha\left[\frac{\kappa}{m^2}+\frac{\kappa^2}{2m^2}\left(1-\frac{t}{4m^2}\right)\right]$ | 0 | $-4\pi\alpha(1+\kappa)$ | $-\frac{\pi\alpha\kappa^2}{2m^4}$ |

For the anomalous magnetic moment $\kappa$ one has to use the values $\kappa_p = 1.79$ and $\kappa_n = 1.91$ for proton and neutron, respectively.

If one considers positive t-values, i.e., time-like momentum transfers, the $A_i(s,t)$ must develop pole-type singularities at the $\pi^0$- and $\eta$-meson masses

$$\sim \frac{1}{t-m_\pi^2} \quad \text{and} \quad \frac{1}{t-m_\eta^2} .$$

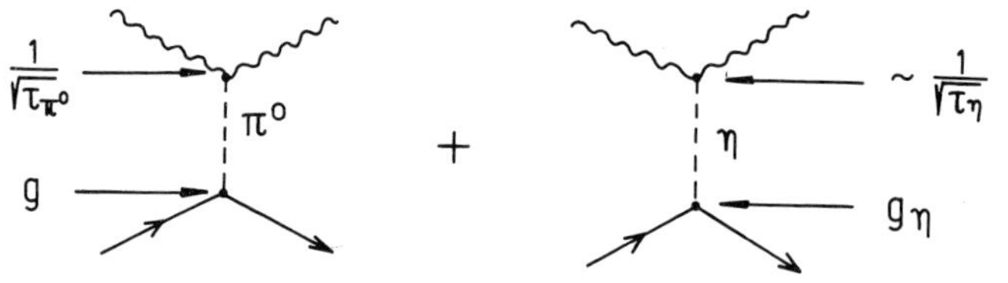


Fig.7 Low diagrams

These are due to the "Low diagrams" of Fig.7. The residues of these meson poles are proportional to

$$g\,\frac{1}{\sqrt{\tau_{\pi^0}}} \quad \text{and} \quad g_\eta\,\frac{1}{\sqrt{\tau_\eta}} ,$$

respectively; $g = \pi N\bar{N}$-coupling constant, $g_\eta = \eta N\bar{N}$-coupling and $\tau_{\pi^0}$ ($\tau_\eta$) are the lifetimes of the $\pi^0$- and the $\eta$-meson.

The corresponding contributions to the $A_i$ we call $A_i^{\pi^0,\eta}$. According to the usual Feynman rules we obtain from Fig.7 and our decomposition (32,41)--compare [30])--

$$A_i^{\pi^0,\eta} = 0 \quad i = 1, \cdots, 5$$

$$A_6^{(\pi^0,\eta)} = \frac{1}{2mm_{(\pi^0,\eta)}} g_{(\pi,\eta)N\bar{N}}\, g_{(\pi,\eta)\gamma\gamma} \frac{1}{m^2_{(\pi^0,\eta)} - t}, \tag{51}$$

where the coupling constants are normalized as follows:

$$g_{\pi N\bar{N}} = 13.4\ , \qquad \Gamma_{(\pi,\eta)\gamma\gamma} = \frac{\pi}{4} m_{(\pi^0,\eta)} \alpha^2 g^2_{(\pi,\eta)\gamma\gamma}\ .$$

Experimentally we have [31]

$$\Gamma_{\pi\gamma\gamma} = (8.02 \pm 0.42)\ \text{eV}$$

and [32]

$$\Gamma_{\eta\gamma\gamma} = (0.324 \pm 0.046)\ \text{keV}\ .$$

One difference compared to pion photoproduction should be observed: the pion pole contribution to the Compton effect is gauge invariant by itself.

## 1.5 Details About the Low-Energy Theorems

In this section we discuss the physical basis of the low-energy properties (3a,b) and give a simple proof for a special case. The essential ingredient is local charge conservation (or gauge invariance) expressed formally by (35). First we sketch Low's argument [3] to understand how low-energy theorems can evolve. In Coulomb gauge ($\varepsilon^0_{(\lambda)} = 0$) only $T_{mn}$ contribute to the Compton amplitude. From (35) one finds

$$k'^m T_{mn} k^n = -k'^0 T_{0n} k^n = k'^0 k^0 T_{00}\ . \tag{52}$$

Here only the retarded commutator of the charge densities enters.

$$T_{00} = -4\pi\alpha i \int d^4x\, e^{ik'\cdot x} \langle p'|\Theta(x_0)[J_0(x),J_0(0)]|p\rangle \ . \qquad (53)$$

Inserting a sum over intermediate states

$$\sum_n \langle p'|J_0(x)|n\rangle\langle n|J_0(0)|p\rangle \qquad (54)$$

one can convince oneself that for low photon energies only the one-nucleon states are important [3], because of the energy gap between the nucleon and the nucleon + pion states. Thus $T_{00}$ will be determined by products of the nucleon matrix elements of the charge density

$$e\langle q|J_0|p\rangle \ .$$

Up to linear terms in $k = q - p$ this is given for example for the proton by

$$\langle q|J_0|p\rangle = \bar{u}(q)\left[\gamma_0 + i\,\frac{\kappa_p}{2m}\,\sigma_{0n}k^n\right]u(p) + O((k^0)^2) \ . \qquad (55)$$

Non-relativistically only the first term remains, leading to the Thomson amplitude $-\alpha/m$. It is by the relativistic contribution of the anomalous magnetic moment to the charge density that $\kappa_p$ comes into play and the Low-Gell-Mann-Goldberger amplitude $-\alpha\kappa_p^2/2m^2$ appears.

A simple formal proof can be given for the Thomson limit (3a) as follows: For $f_1$ we can disregard the nucleon spin. Therefore, the tensor $T_{\mu\nu}(k,p)$ for the Compton forward scattering ($p = p'$, $k = k'$ !) can be written by Lorentz invariance as

$$T_{\mu\nu} = a_1 g_{\mu\nu} + a_2(p_\mu k_\nu + k_\mu p_\nu) + a_3(p_\mu k_\nu - k_\mu p_\nu) + a_4 p_\mu p_\nu + a_5 k_\mu k_\nu \ .$$

From the gauge invariance conditions $k'^\mu T_{\mu\nu} = T_{\mu\nu}k^\nu = 0$ one finds that $a_3$ and $a_4$ must vanish, $a_5$ can be arbitrary and $a_1$ and $a_2$ must obey the constraint

$$a_1 + m\nu a_2 = 0 \ . \qquad (57)$$

This is the essential gauge condition which allows us to deduce an exact expression for $a_1$ ($\nu = 0$) if we use also crossing and analyticity. Crossing requires

$$a_1(-\nu) = a_1(\nu) \; ; \quad a_2(-\nu) = -a_2(\nu) \; . \tag{58}$$

To exploit analyticity we write

$$a_i(\nu) = a_i^{\text{Pole}}(\nu) + \bar{a}_i(\nu) \tag{59}$$

where the first term contains the nucleon singularities corresponding to Fig.6, while the remainder $\bar{a}_i$ is regular at $s = M^2$ and $u = M^2$. From (58) one finds for the pole terms--using $s - M^2 = 2M\nu$, $u - M^2 = -2M\nu$--

$$a_1^{\text{Pole}} \sim \left(\frac{1}{s - m^2} + \frac{1}{u - m^2}\right) = 0 \; , \quad a_2^{\text{Pole}} = C\left(\frac{1}{s - m^2} - \frac{1}{u - m^2}\right) = \frac{C}{m\nu} \; .$$

The constant C can be found to be

$$C = \alpha$$

by explicit evaluation of the Born approximation. Using now (57) and the regularity of $\bar{a}_2$ at $\nu = 0$ we find

$$a_1(0) = -\alpha - \lim_{\nu \to 0} (m\nu\bar{a}_2(\nu)) = -\alpha \; . \tag{60}$$

This is just the Thomson limit (3a) if the kinematic factor W between $f_1$ and $a_1$ is taken into account.

For the general case of non-forward Compton scattering off nucleons we start with the s-channel helicity amplitudes. According to (43) the $\Phi_k$ have the following structure

$$\Phi_k(s, \cos\Theta) = \frac{(s - m^2)^2}{s^2} \sum_{i=1}^{6} p_{ki}(\sqrt{s}, \cos\Theta) A_i(s,t) \tag{61}$$

where the $p_{ki}$ are known polynomials with respect to $\sqrt{s}$ and $\cos\Theta$.

From analyticity--expressed in the form of a dispersion relation in (49)--we know that

$$A_i(s,t) = A_i^{\text{Born}}(s,t) + A_i^{C}(s,t) \tag{62}$$

where the continuum contribution $A_i^C$ is finite (or zero) at $s = m^2$ or $u = m^2$.

Therefore, in an expansion of the helicity amplitudes in powers of $(s-m^2)$ at fixed CMS-angle $\Theta$ only the Born term contributes to the zeroth- and first-order terms. Due to the polynomial structure of the coefficients in (61) our statement can most easily be seen by considering

$$(s-m^2)(u-m^2)A_i(s,t)\Big|_{\substack{s\to m^2\\ \Theta\ \text{fixed}}} \quad . \tag{63}$$

By means of (62) and the explicit form of the Born terms (Eq.(50)) we obtain for (63)

$$R_i^{su} + 2R_i^-(0)(s-m^2) - \left[A_i^C(m^2,0) + C_i - \frac{1}{2m^2}(1-\cos\Theta)(R_i^+(0)+R_i^-(0))\right]$$

$$\times\ (s-m^2)^2 + O((s-m^2)^3)$$

where $A_i^C(m^2,0)=0$ for the crossing odd amplitudes.

If we put the corresponding expansion of the $\Phi_k(s,\cos\Theta)$ into our cross-section formula (25) we get the differential cross section up to the second power in the CMS momentum expressed in terms of the electric charge e, the anomalous magnetic moment $\kappa$ and two appropriate combinations of the constants $A_i^C(m^2,0)$ (compare [25]).

Explicitly we obtain, including terms up to first order in the CMS momentum

$$T = \frac{4\pi\alpha}{m}\,\underset{\sim}{\varepsilon}'\cdot\underset{\sim}{\varepsilon} + i\,\frac{4\pi\alpha}{m^2}\,\omega\underset{\sim}{\sigma}\cdot(\underset{\sim}{\varepsilon}\times\underset{\sim}{\varepsilon}') + \frac{2\pi i\alpha}{m^2}(1+\kappa)^2\omega^{-1}\underset{\sim}{\sigma}\cdot[(\underset{\sim}{k}'\times\underset{\sim}{\varepsilon}')\times(\underset{\sim}{k}\times\underset{\sim}{\varepsilon})]$$

$$+ \frac{2\pi i\alpha}{m^2}(1+\kappa)\,\frac{1}{2\omega}\,[(\underset{\sim}{\varepsilon}\cdot\underset{\sim}{k}')(\underset{\sim}{\varepsilon}'\cdot\underset{\sim}{\sigma}\times\underset{\sim}{k}') - (\underset{\sim}{\varepsilon}'\cdot\underset{\sim}{k})(\underset{\sim}{\varepsilon}\cdot\underset{\sim}{\sigma}\times\underset{\sim}{k})] + O(\omega^2)\ . \tag{64}$$

This expression is written in the barycentric system; $\underset{\sim}{k}$, $\underset{\sim}{k}'$ and $\omega$ are the CMS-photon momenta and the CMS-photon energy. The polarization vectors $\underset{\sim}{\varepsilon}$ and $\underset{\sim}{\varepsilon}'$ fulfill the Coulomb gauge condition $\underset{\sim}{\varepsilon}\cdot\underset{\sim}{k}=\underset{\sim}{\varepsilon}'\cdot\underset{\sim}{k}'=0$.

## 1.6 Experimental Data in the Low-Energy and Resonance Region

From (64) one finds for $\nu\to 0$

$$\frac{d\sigma}{d\Omega} = \frac{1}{2} \sum_{\nu,\nu'=\pm 1/2} \frac{1}{2} \sum_{\lambda,\lambda'=\pm 1} \left| \frac{\alpha}{m} \underset{\sim}{\varepsilon}(\lambda') \cdot \underset{\sim}{\varepsilon}(\lambda) \right|^2 \tag{65}$$

i.e.,

$$\frac{d\sigma}{d\Omega} = \left(\frac{\alpha}{m}\right)^2 \frac{1 + \cos^2\Theta}{2} ,$$

where the transversality of the photon is responsible for the non-trivial angular dependence. On the other hand, a multipole expansion (compare Table 1) with only the lowest amplitudes $f_{EE}^{1\pm}$ gives a differential cross section of the form

$$\frac{d\sigma}{d\Omega} = 2\left|f_{EE}^{1-}\right|^2 + \frac{1}{2}\left|f_{EE}^{1+}\right|^2 (3 \cos^2\Theta + 7) + \mathrm{Re} \left\{ \left( f_{EE}^{1-*} f_{EE}^{1+} \right) (3 \cos^2\Theta - 1) \right\} . \tag{66}$$

This result can be conveniently read off from Table 4 which is taken from [33]. Therefore the Thomson limit corresponds to

$$f_{EE}^{1-} = f_{EE}^{1+} = - \frac{1}{3} \frac{\alpha}{m} , \tag{67}$$

i.e., the virtual intermediate states with spin-parity $J^P = 1^-/2$ and $3^-/2$ are excited and de-excited with the same probability amplitude.

We get a feeling for the order of magnitude of Compton scattering cross section if we use in (65) the value for

$$\frac{\alpha}{m} \simeq 1.5 \times 10^{-16} \text{ cm} .$$

Any theoretical approach to Compton scattering incorporates (65). Unfortunately experimental data are available only about photon energies $E_\gamma^{Lab} \geq 50$ MeV. Since the energy dependence between 50 and 150 MeV is nearly constant, these data, however, give a qualitative successful test of the $1 + \cos^2\Theta$-law; compare [34].

At higher energies one expects the increasing influence of the $\Delta$ (1236)-isobar. From photoproduction we know that $\Delta$ is excited mainly by a magnetic dipole, thus the amplitude $f_{MM}^{1+}$ is expected to dominate around 320 MeV. The corresponding angular dependence is, according to Table 4, given by

Table 4 Angular distribution for multipoles with $L \leq 2$ ($x = \cos\Theta$) [33]

| | $f_{EE}^{1-}$ | $f_{MM}^{1-}$ | $f_{EE}^{1+}$ | $f_{MM}^{1+}$ | $f_{EM}^{1+}$ |
|---|---|---|---|---|---|
| $f_{EE}^{1-}$ $S_{1/2}$ | 2 | 4x | $3x^2 - 1$ | 2x | $-6(3x^2 - 1)$ |
| $f_{MM}^{1-}$ $P_{1/2}$ | | 2 | 2x | $3x^2 - 1$ | -12x |
| $f_{EE}^{1+}$ $D_{3/2}$ | | | $(1/2)(3x^2 + 7)$ | 10x | $6(3x^2 - 1)$ |
| $f_{MM}^{1+}$ $P_{3/2}$ | | | | $(1/2)(3x^2 + 7)$ | 12x |
| $f_{EM}^{1+}$ $D_{3/2}$ | | | | | $18(1 + x^2)$ |
| $f_{ME}^{1+}$ $P_{3/2}$ | | | | | |
| $f_{EE}^{2-}$ $P_{3/2}$ | | | | | |
| $f_{MM}^{2-}$ $D_{3/2}$ | | | | | |
| $f_{EE}^{2+}$ $F_{5/2}$ | | | | | |

| | $f_{ME}^{1+}$ | $f_{EE}^{2-}$ | $f_{MM}^{2-}$ | $f_{EE}^{2+}$ |
|---|---|---|---|---|
| $f_{EE}^{1-}$ $S_{1/2}$ | -12x | 18x | $9(3x^2 - 1)$ | $6x(5x^2 - 3)$ |
| $f_{MM}^{1-}$ $P_{1/2}$ | $-6(3x^2 - 1)$ | $9(3x^2 - 1)$ | 18x | $6(3x^2 - 1)$ |
| $f_{EE}^{1+}$ $D_{3/2}$ | 12x | $18x(2x^2 - 1)$ | $9(3x^2 - 1)$ | $6x(4x^2 + 3)$ |
| $f_{MM}^{1+}$ $P_{3/2}$ | $6(3x^2 - 1)$ | $9(3x^2 - 1)$ | $18x(2x^2 - 1)$ | $21(3x^2 - 1)$ |
| $f_{EM}^{1+}$ $D_{3/2}$ | $72x^3$ | $-36x(4x^2 - 3)$ | $-18(3x^2 - 1)$ | $36x(4x^2 - 3)$ |
| $f_{ME}^{1+}$ $P_{3/2}$ | $18(1 + x^2)$ | $-18(3x^2 - 1)$ | $-36x(4x^2 - 3)$ | $18(3x^2 - 1)$ |
| $f_{EE}^{2-}$ $P_{3/2}$ | | $(9/2)(3x^2 + 7)$ | $-54x + 144x^3$ | $9(40x^4 - 33x^2 + 3)$ |
| $f_{MM}^{2-}$ $D_{3/2}$ | | | $(9/2)(3x^2 + 7)$ | $18x(12x^2 - 7)$ |
| $f_{EE}^{2+}$ $F_{5/2}$ | | | | $18(5x^4 - 3x^2 + 3)$ |

$$|f_{MM}^{1+}|^2 \, \frac{1}{2} \, (3 \cos^2\Theta + 7) \; . \qquad (68)$$

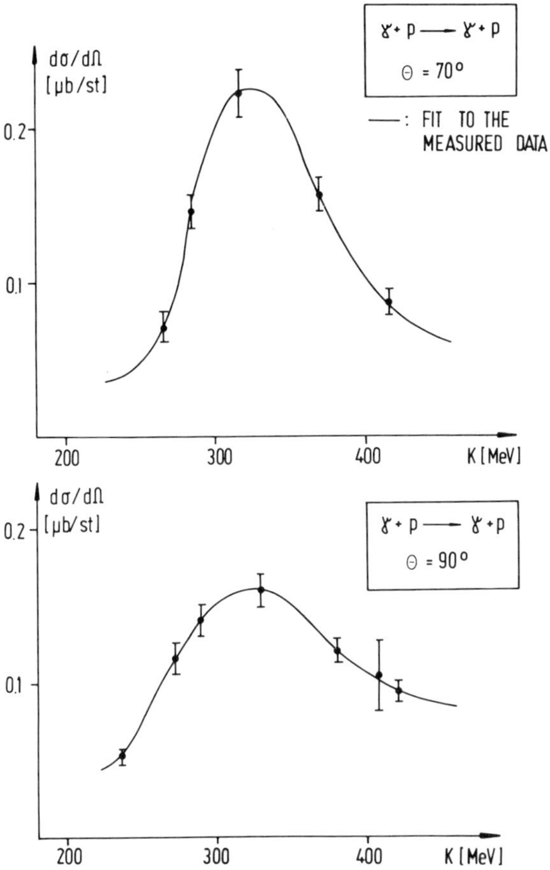


Fig.8 Energy dependence of $d\sigma/d\Omega$ at $\Theta = 70°$ and $90°$ [35]

The experimental excitation functions indeed exhibit a pronounced resonance structure; see Fig.8, which shows the results of a recent experiment [35]. It is amusing that the value of the total cross section at the resonance energy $E_\gamma^{Lab} = 320$ MeV can be calculated from the simple Breit-Wigner formula

$$\sigma_{tot} = \left(J + \frac{1}{2}\right) \frac{2\pi}{k^2} \frac{\Gamma_\gamma^2}{\Gamma^2}$$

using the usual total width of the $\Delta$ (1236) and a partial photon width

$$\Gamma_\gamma = 0.72 \pm 0.03 \text{ MeV} \; ,$$

in complete agreement with $\Gamma_\gamma = 0.69 \pm 0.02$ MeV found in photoproduction (see [36]). But the observed angular distribution is very far from (68). A quadratic fit

$$\frac{d\sigma}{d\Omega} = A + B\cos\Theta + C\cos^2\Theta$$

gives values of A/C between 1.2 and 0.75 (in contrast to 2.3 from (68)). Also, a significant asymmetry term B is present, B/C ranging from 0.1 to 0.4, indicating an appreciable interference [35]. Also theory has to work hard to understand the new data. See Fig.9 where all data between 200 and 400 MeV are shown, together with a dispersion theoretical calculation [38]. The new data (ǂ) have a resonance at the expected position 320 MeV but are smaller than the older values and the dispersion theory.

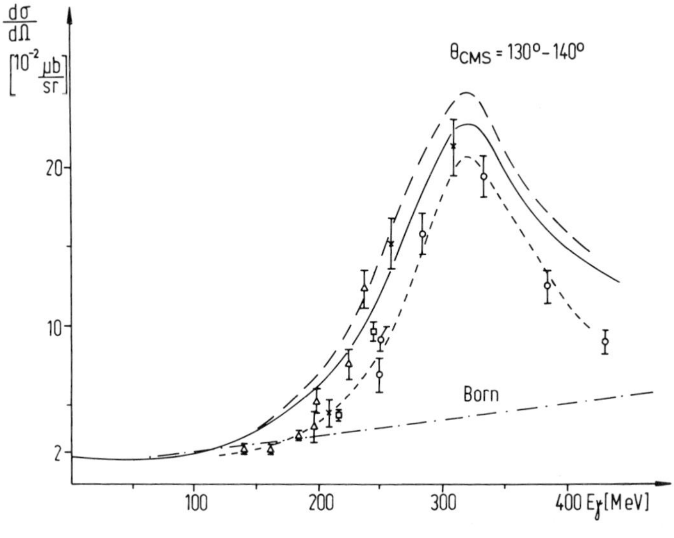


Fig.9 Excitation curve dσ/dΩ(135°). Data from [35,37]. The curves represent different dispersion calculations (compare [38])

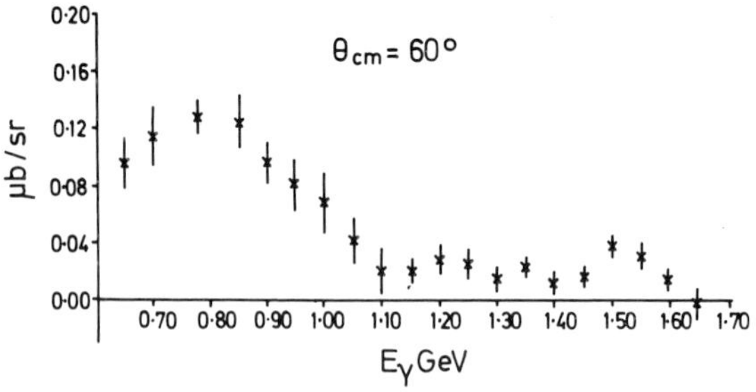


Fig.10 Energy dependence of dσ/dΩ (60°) between 0.6 and 1.7 GeV [39]

An impression about the Compton cross section between 600 MeV and 1.7 GeV is given in Fig.10 based on data from [39]. The $D_{13}$ (1520) and $F_{37}$ (1890) isobars are clearly exhibited by the data as a 2nd and 4th resonance at 0.8 and 1.5 GeV while the indicated flat maximum at 1.2 GeV lies a little too high for the $F_{15}$ (1688) isobar.[7] Further and more accurate data are needed.

# 2. High-Energy Compton Scattering

## 2.1 Gross Features of the Experimental Data

The experiments performed during the last couple of years at different laboratories [9,40-42] show the following gross features:

i) The total hadronic cross section $\sigma_{tot}^{\gamma N}$--related through the optical theorem to the imaginary part of the forward scattering amplitude--tends at high energy to a constant, isospin independent value. The data allow a fit of the form (compare Fig.1)

$$\sigma_{tot}^{\gamma N} = C_p + C_{I=0}E_\gamma^{-1/2} + C_{I=1}E_\gamma^{-1/2} \tag{69}$$

with [9]

$$C_p = 97.4 \pm 1.9\ \mu b$$

$$C_{I=0} = 55.0 \pm 5.1\ \mu b$$

$$C_{I=1} = 12.3 \pm 2.3\ \mu b\ .$$

Data taken by CALDWELL et al. [42] for $\sigma_{tot}^{\gamma p}$ from 4 to 18 GeV are in agreement with (69). Slightly different values for $C_p$ and $C_{I=0} + C_{I=1}$ have been reported by ARMSTRONG et al. [41]. Other fits with different exponents of $E_\gamma$ in (69) (ranging between -0.4 and -0.6) are possible [41].

ii) The real part of the no spin-flip forward scattering amplitude $f_1(\nu)$ may be computed by means of the dispersion relation (12) and the experimental data on $\sigma_{tot}^{\gamma N}$. The result of such a calculation is shown in Fig.11.

It turns out that the ratio Real $f_1$/Im $f_1$ tends to zero at high energies.

---

[7] Compare also the data from [40].

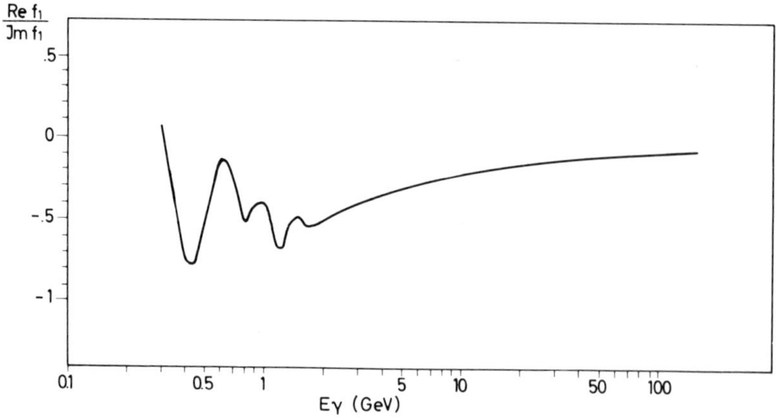


Fig.11 Ratio of the real to imaginary part of $f_1$ for Compton scattering on protons (from [9])

iii) The differential cross section $d\sigma/dt$ shows an exponential fall off with increasing momentum transfer t at fixed energy. A fit of the form

$$\frac{d\sigma}{dt} = \frac{d\sigma^0}{dt} \exp(At + Bt^2) \tag{70}$$

has been performed for the data in the t-range $|t| \lesssim 1\ \mathrm{GeV}^2$ at different energies. The resulting values of A and B are listed in Table 5.

For comparison we note the slope parameters for $\pi p$-scattering at 9 GeV and $|t| < 1\ \mathrm{GeV}^2$ [9]

$$A_{\pi p} = 9.0 \pm 0.2\ \mathrm{GeV}^{-2}$$

$$B_{\pi p} = 2.5 \pm 0.3\ \mathrm{GeV}^{-4} \ .$$

iv) The data are consistent with the assumption of s-channel helicity conservation (SCHC), i.e.,

$$M_{\lambda',\nu';\lambda,\nu} = 0 \quad \text{if} \quad \lambda \neq \lambda' \quad \text{and(or)} \quad \nu \neq \nu' \ . \tag{71}$$

a) With our knowledge of $\mathrm{Re}\{f_1\}/\mathrm{Im}\{f_1\}$ from ii) and the assumption of a vanishing spin-flip amplitude for forward scattering ($f_2 = 0$) we obtain from the optical theorem

$$\frac{d\sigma^0}{dt} = \frac{\sigma^2_{tot}}{16\pi} \left[1 + \left(\frac{\mathrm{Re}\{f_1\}}{\mathrm{Im}\{f_1\}}\right)^2\right] \ . \tag{72}$$

Table 5 Fit of $d\sigma/dt$ for scattering on protons to the forms $d\sigma^0/dt \exp(At)$ and $d\sigma^0/dt \exp(At + Bt^2)$, respectively. (Data with an asterisk are from [40]; all others are from [9])

| Experiment | E (GeV) | t range ($GeV^2$) | A ($GeV^{-2}$) | B ($GeV^{-4}$) |
|---|---|---|---|---|
| MIT | 2.0±0.5 | 0.14-0.41 | 5.3±0.5* | - |
| DESY | 2.2-2.7 | 0.1 -0.4 | 5.2±0.5 | - |
| | 2.7-3.2 | - | 5.7±0.4 | - |
| MIT | 3.2±0.6 | 0.14-0.41 | 4.6±0.4* | - |
| DESY | 3.2-3.7 | - | 6.2±0.4 | - |
| | 3.7-4.2 | - | 5.3±0.5 | - |
| | 4.0-5.2 | 0.06-0.4 | 6.0±0.4 | - |
| | 5.0-6.2 | - | 5.5±0.3 | - |
| | 6.0-7.0 | - | 5.9±0.4 | - |
| SLAC | 8 | 0.014-0.17 | 7.7±0.5 | - |
| | | 0.014-0.8 | 7.6±0.4 | 2.3±0.5 |
| | 16 | 0.014-0.17 | 7.9±0.5 | - |
| | | 0.014-1.1 | 7.3±0.3 | 1.7±0.3 |

Both sides of (72) are compared with each other in Fig.12. From the data we conclude that $f_2$ contributes at most 10% to $d\sigma^0/dt$.

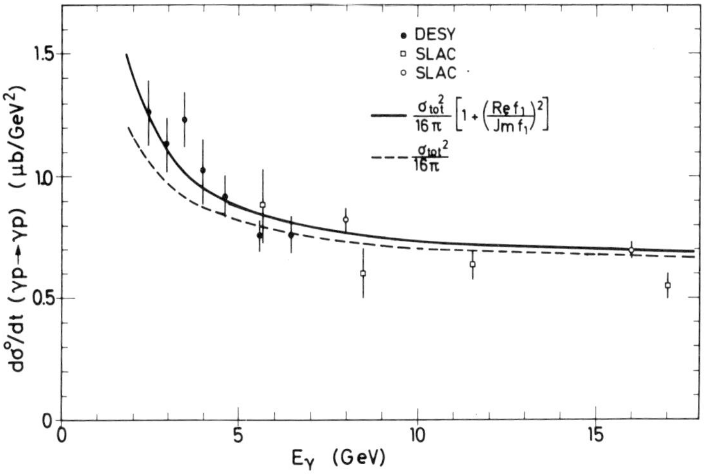


Fig.12 Extrapolated forward cross section of proton Compton scattering as a function of the photon energy $E_\gamma$ (from [9])

b) The photon asymmetry $\Sigma$ defined by (26) has been measured at 3.5 GeV and different t-values at DESY [43]. The observed vanishing of $\Sigma$ within experimental errors is, according to (26), consistent with SCHC (compare Fig.13).

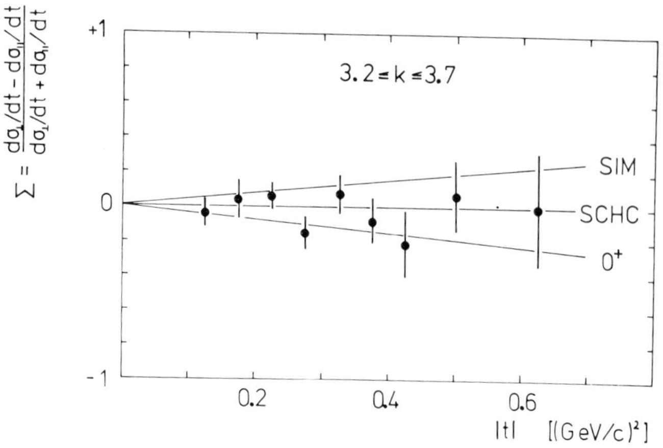


Fig.13 Asymmetry $\Sigma$ of Compton scattering on protons vs. four momentum transfer. Solid lines are the predictions of the spin independent model (SIM), s-channel helicity conservation and $0^+$ exchange in the t-channel (from [43])

The properties of the Compton scattering amplitude collected in i) to iv) are typical for the behavior of an elastic hadron-hadron scattering amplitude at high energies. Therefore in nucleon Compton scattering at high energies, the photon shows a hadron-like behavior, which quantitatively can be expressed by a direct proportionality between the total cross sections $\sigma^T_{\gamma N}$ and $\sigma^T_{\pi N}$ [44]

$$\sigma^T_{\gamma N} \simeq \frac{1}{220}\,\sigma^T_{\pi^0 p} \; . \tag{73}$$

This relation may be explained by means of the vector dominance model which relates photon amplitudes to the corresponding vector meson amplitudes [45].

Relation (73) is also valid in the first resonance region but around the second resonance the photon cross section is notably larger. This fact has been interpreted as due to couplings to higher spin particles, the so-called contact terms [46,47].

## 2.2 Regge Model

The data on two-body hadronic reactions at high energies and low momentum transfer are best described in terms of a Regge pole model. The hadron-like behavior of photons in high-energy Compton scattering found in Section 2.1 suggests that we apply the Regge picture in this case too. The contribution of a single Regge pole to Compton scattering is illustrated in Fig.14.

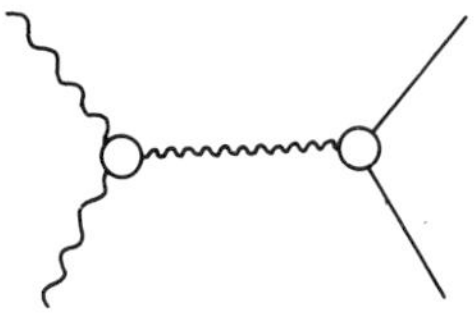

Fig.14 Generalized Feynman diagram for the contribution of a single Regge pole

The quantum numbers of Reggeons contributing to Compton scattering are determined by the conservation laws at the $N\bar{N}R$ and $\gamma\gamma R$ vertices, respectively. Hence R must possess the quantum numbers

$$I = 0,1 \ , \quad C = 1 \ , \quad \text{and therefore} \quad G = (-1)^I \ .$$

Admissible Reggeons with natural parity are P, P', $\varepsilon$, $\delta$, S*, $A_2$, and its daughters, those with unnatural parity, are $\pi_0$, B, $\eta$, $\eta'$, $A_1$, D, $A_3$.

Possible effects of fixed poles will be discussed later. Let us first try to find out whether we are able to explain the main features of the experimental data by means of moving poles only.

A common feature of elastic hadron scattering is the dominance of the *Pomeron (P)-exchange* at high energies. In particular, P-exchange guarantees the constancy of total cross sections, due to $\alpha_P(0) = 1$.

But Regge-exchange with an intercept equal to one at $t = 0$ is forbidden for the forward helicity non-flip Compton amplitude $f_1$ due to angular momentum conservation, as can be seen from the following simple argument: Choose the momentum of the incoming photon $\underset{\sim}{k}$ as quantization axis. Then for $t = 0$ conservation of the 3-component of angular momentum yields (for notation compare Fig.4)

$$\lambda - \nu = \lambda' - \nu' \ .$$

Because $\lambda(')$ only takes values $\pm 1$ and $\nu(')$ is restricted to $\pm 1/2$ we must have

$$\lambda - \lambda' = 0 \ . \tag{74}$$

If we turn to the t-channel amplitude, the crossing relations for photon helicities are of a particularly simple form due to the zero mass of the photon [48]. The photon helicity changes sign if an incoming photon becomes an outgoing one or vice versa. Therefore we have instead of (75)

$$\lambda_t + \lambda_t' = 0 \ , \tag{75}$$

i.e., the 3-component of angular momentum in the t-channel $\lambda_t - \lambda'_t = 2\lambda_t$ must be $\pm 2$ for $t = 0$ which forbids total angular momentum equal to one.

In order to give a more formal treatment of this argument, we consider the simpler case of Compton scattering off scalar nucleons.

Equation (41) tells us that in the scalar case only the covariants $I^{1,2}_{\mu,\nu}$ survive. Therefore, only the invariant amplitudes $A_{1,2}$ have to be taken into account.

From the two t-channel helicity amplitudes $f^t_{1,\pm 1}$ the non-flip amplitude $f^t_{1,1}$ vanishes at $t = 0$ according to (45). The remaining flip amplitude $f^t_{1,-1}$ looks to leading order in s as follows:

$$\left(\text{notice:} \quad \cos\Theta_t \underset{s\to\infty}{\to} \frac{2s}{\sqrt{t(t-4m^2)}}\right) \quad , \qquad f^t_{1,-1} \simeq -\frac{s^2}{2}\sqrt{\frac{t}{4} - m^2}\, A_2(s,t) \; . \tag{76}$$

From (73) we read off that i) a Regge-contribution with $\alpha(0) = 1$ must vanish at $t = 0$ due to the arguments given above; ii) $f^t_{1,-1}$ must be analytic in a neighborhood of $t = 0$ as $A_2$ does not possess any dynamical singularity there.

From i) and ii) we conclude that the Pomeron contribution to $f^t_{1,-1}$ (which appears only in the even signature part $f^{t(+)}_{1,-1}$) vanishes at $t = 0$ like a power of $\alpha_p(t) - 1$, i.e.,

$$f^{tP}_{1,-1} \underset{t\to 0}{\sim} (\alpha_p(t) - 1)^n \beta^0_p(t) s^{\alpha_p(t)} \tag{77}$$

with $\beta^0_p(t)$ being a reduced Pomeron residue. In order to localize the origin of such a behavior, thereby trying to determine n, we consider the partial wave expansion of $f^{t(+)}$

$$f^{t(+)}_{1,-1} = \sum_{J=2}^{\infty} d^J_{20}(\cos\Theta_t)(2J+1)\, \frac{1}{2}\, (1 + e^{-i\pi J}) a^{(+)}(J,t) \; . \tag{78}$$

The angular momentum $J = 1$ does not occur in (78); thus $J = 1$ is a nonsense point for the amplitude $a^{(+)}(J,t)$ (compare [49]). Moreover, the signature factor $1 + e^{i\pi J}$ vanishes for $J = 1$; therefore $J = 1$ is also a wrong signature point.

As one square root $[\alpha_p(t) - 1]^{1/2}$ is included already in the analytic continuation of $d^J_{20}$ [48], a power $n - 1/2$ must be contained in the Pomeron residue of the analytic continuation of $a^{(+)}(J,t)$

$$\beta_p(t) = [\alpha_p(t) - 1]^{n-1/2} \beta_p^0(t) \ . \tag{79}$$

Therefore, the natural number n is a phenomenological parameter. We choose the simplest possibility $n = 1$ (for a discussion of other choices compare [49]).

Equation (77) would lead to a vanishing total cross section. This contradiction to experiment could be avoided if the reduced residue $\beta_p^0(t)$ happens to be singular at $t = 0$. According to general Regge theory this can be achieved by means of a fixed pole at $J = 1$, which couples multiplicative to the Pomeron and leads to an expression[8]

$$\beta_p^0(t) = \frac{1}{\alpha_p(t) - 1} \beta_p^{00}(t) \tag{80}$$

where $\beta_p^{00}(t)$ is regular at $t = 0$.

Let us go back to Compton scattering off spin 1/2-nucleons. According to Reggeology (see [48]) a single Regge-pole gives a contribution $\sim s^\alpha$ at $s \to \infty$ to the t-channel helicity amplitudes $f^t_{\lambda',\nu';\lambda,\nu}$. This implies, due to the kinematical factors in (45), $A_{1,6} \sim s^\alpha$, $A_5 \sim s^{\alpha-1}$, and $A_{2,3,4} \sim s^{\alpha-2}$ for $s \to \infty$. Therefore, we obtain for the s-channel helicity amplitudes from (43) to leading order in s [23]

$$f^s_{1\frac{1}{2},1\frac{1}{2}} \simeq -s^2 \left(\frac{m}{2} A_2 + A_3 - A_4\right) \ , \quad f^s_{1-\frac{1}{2},1-\frac{1}{2}} \simeq -s^2 \left(\frac{m}{2} A_2 + A_3 + A_4\right) \ ,$$

$$f^s_{1\frac{1}{2},1-\frac{1}{2}} \simeq \frac{1}{4} s^2\sqrt{-t}\, A_2 \ , \quad f^s_{1\frac{1}{2},-1\frac{1}{2}} \simeq mtA_1 \ , \tag{81}$$

$$f^s_{1\frac{1}{2},-1-\frac{1}{2}} \simeq -\frac{1}{2}\sqrt{-t}\,(tA_1 + 8sA_5 + 2mtA_6) \ , \quad f^s_{1-\frac{1}{2},-1\frac{1}{2}} \simeq \frac{1}{2} t\sqrt{-t}\,(A_1 - 2mA_6) \ .$$

From *s-channel helicity conservation* at high energy (compare Sec.2.1) we then conclude

$$A_1 = s^2 A_2 = sA_5 = A_6 = 0 \tag{82}$$

to leading order in s.

---

[8] Details are discussed in the Appendix; compare especially (A.7).

In terms of the remaining amplitudes $A_3$ and $A_4$ the t-channel helicity amplitudes take at high energies, according to (45), the following form

$$f^t_{1,-1;\frac12,\frac12} \simeq \frac{s^2 m}{p} A_3 \ , \quad f^t_{1,-1;-\frac12,\frac12} \simeq -(\sin\Theta_t)\, \frac{s}{p} \sqrt{t}\, (-qA_3 + pA_4) \ ,$$

$$f^t_{1,-1;\frac12,-\frac12} \simeq -(\sin\Theta_t)\, \frac{s}{p} \sqrt{t}\, (qA_3 + pA_4) \ , \quad f^t_{1,1;\frac12,\frac12} \simeq \frac{p(p+q)}{m} \sqrt{t}\, (p-q)\, \frac{s}{p} A_4 \ ,$$

$$f^t_{1,1;-\frac12,-\frac12} \simeq \frac{p(q-p)}{m} \sqrt{t}\, (p+q)\, \frac{s}{p} A_4 \ , \quad f^t_{1,1;-\frac12,\frac12} \simeq m^2 t (\sin\Theta_t) A_4 \ . \tag{83}$$

From (83) we conclude that $A_4$ contributes at high energy to unnatural parity-exchange in the t-channel only. The $N\bar{N}$-angular momentum eigenstates in the helicity basis transform under parity as follows [22]:

$$P|J,M;\nu,\nu'\rangle = (-1)^{J-\nu-\nu'+1}|J,M;-\nu-\nu'\rangle \ . \tag{84}$$

Then the combination

$$|J,M;1/2,-1/2\rangle + |J,M;-1/2,1/2\rangle$$

is a parity eigenstate with eigenvalue $(-1)^{J+1}$ (unnatural parity).

Therefore, the combination of t-channel helicity amplitudes

$$f^t_{1,-1;\frac12,-\frac12} + f^t_{1,-1;-\frac12,\frac12} \simeq -2(\sin\Theta_t) s\sqrt{t}\, A_4 \tag{85}$$

contains unnatural parity exchange in the t-channel only. The $t = 0$ intercepts of Regge trajectories with unnatural parity are much lower than those with natural parity. Hence, due to the common belief of parallel Regge trajectories, we might neglect all unnatural parity Reggeons.

For the surviving invariant amplitude $A_3$ we make the usual Regge Ansatz[9]

[9] Note that for small t and linear trajectories we have

$$\frac{\frac12 [1 + e^{-i\pi\alpha(t)}]}{\sin\pi\alpha(t)} = \frac{e^{-i\frac{\pi}{2}\alpha(t)}}{2\sin\frac{\pi}{2}\alpha(t)} \simeq \frac{e^{-i\frac{\pi}{2}\alpha(t)}}{2\sin\frac{\pi}{2}\alpha(0)} \ .$$

$$A_3^{Regge}(\nu,t) = -\sum_i \beta_i(t)\, e^{-i(\pi/2)\alpha_i(t)} \left(\frac{2\nu m}{m_0^2}\right)^{\alpha_i(t)-2} \tag{86}$$

where the sum runs over P, P', $A_2$ and its first daughter[10] $A_2'$.

In terms of $A_3$ the cross section formulas at sufficiently high energy look as follows:

$$\sigma_{tot}^{\gamma N} \simeq 2m\nu \, \mathrm{Im}\{A_3(\nu,0)\} \tag{87}$$

$$\frac{d\sigma}{dt} \simeq \frac{s^3}{64\pi k^2} \, |A_3(\nu,t)|^2 \; . \tag{88}$$

New data for $\sigma_{\gamma N}^{T}$ and $d\sigma/dt$ have been fitted by KELLET [23] by means of Ansatz (86). He obtained the following results:

Trajectories

$$\alpha_p(t) = 1 + 0.5t$$

$$\alpha_{p'}(t) = 0.5 + 0.9t$$

$$\alpha_{A_2}(t) = 0.3 + 0.9t$$

$$\alpha_{A_2'}(t) = \alpha_{A_2}(t) - 1 \; .$$

$\alpha_p$ and $\alpha_{p'}$ have been taken from fits to hadronic reactions. It turns out that $\sigma_{tot}^{\gamma N}$ is not sensitive with respect to $\alpha_{A_2}(0)$. The value 0.3 has been obtained by means of a fit to the D/H ratio in deep inelastic electron scattering. This is in agreement with BARGER's fit to $\pi N$, KN, $\bar{K}N$, NN and $\pi N \to \eta N$ reactions --he got $0.34 \pm 0.03$ [51].

We treat the Pomeron as an effective single Regge pole, thereby disregarding its complicated nature. In this sense we have to understand also the unusually slow rise of the trajectory.

---

[10] For a fit to real Compton scattering the inclusion of the $A_2'$ daughter is not necessary, but we need it for virtual Compton scattering in the deep inelastic region. The importance of daughters for Regge theory is discussed, for example, in [48].

Residua

The P', $A_2$, $A_2'$ residua have been taken to be constant, whereas the P-residuum has an exponential t-dependence $\beta_P(t) = e^{b_P t}\,\beta_P(0)$.

$$\beta_P(0) = 0.25\ \mathrm{GeV}^{-4}\ , \quad \beta_{P'} = 0.23\ \mathrm{GeV}^{-4}\ , \quad \beta_{A_2} = 0.075\ \mathrm{GeV}^{-4}\ ,$$

$$\beta_{A_2'} = -0.23\ \mathrm{GeV}^{-4}\ , \quad b_P = 1.76\ \mathrm{GeV}^{-2}\ , \quad m_0 = 1.0\ \mathrm{GeV}\ .$$

In Figs.15 and 16 the fit is compared with the experimental data on $\sigma_{tot}^{\gamma N}$ and $d\sigma/dt$, respectively. This fit is not unique.

i) Due to $\alpha_{A_2}(0) \neq 0.5$ this fit for $\sigma_{tot}^{\gamma N}$ is different from (69).

ii) Other authors obtain good agreement with essentially the same data with the assumption of P'-$A_2$ exchange degeneracy ($\alpha_{A_2,P'}(0) = 1/2$) but $\alpha_P'(0) = 0.3\ (\mathrm{GeV})^{-2}/c$ [52].

We conclude that the main features of the data can be explained by means of a pure moving pole Ansatz without cuts and fixed poles.[11]

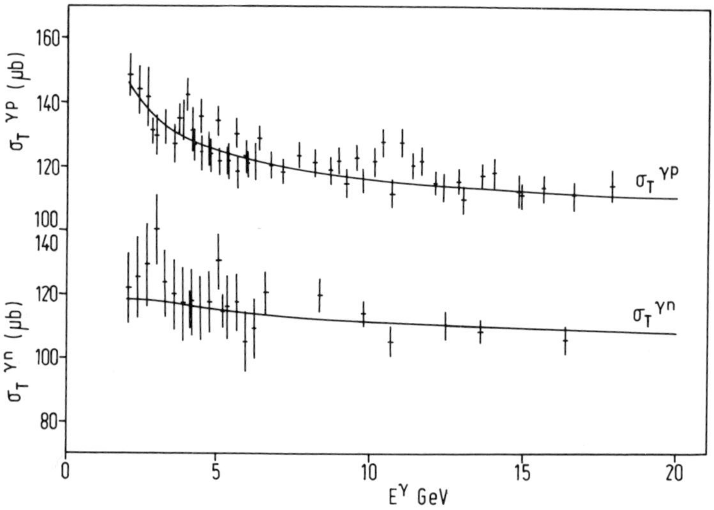


Fig.15 Fit to $\sigma_{tot}^{\gamma N}$ for $E^\gamma > 2$ GeV (from [23])

[11] Remember that the nonsense wrong signature J = 1 fixed pole does not show up explicitly in the cross section. Its only purpose is to reinstate the Pomeron.

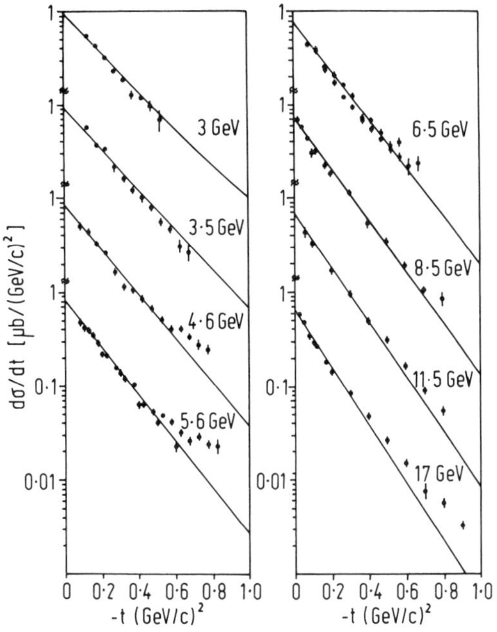


Fig.16 Fit to dσ/dt for different energies between 3 and 17 GeV (from [23])

## 2.3 Fixed Poles

In the following we will consider right-signature fixed poles at $J = 0$ and 1, respectively. Such a possibility has been first discussed by CREUTZ et al. [53].

From the finiteness of partial wave expansions of t-channel helicity amplitudes in the t-channel physical region it follows immediately that right-signature fixed poles at integer J-values are admissible at nonsense points only.

The mechanism allowing right-signature fixed poles in the Compton amplitude is exactly the same as the one described for the wrong signature case in the Appendix. But we must notice one important difference: the signature factor $[1/2(1 \pm e^{-i\pi J})]/\sin\pi J$ becomes a pole near a right-signature integer. Therefore, right-signature fixed poles at nonsense integers will show up explicitly in the analogue of (A.7) and will influence the high-energy behavior of the corresponding amplitude. In particular, a $J^P = 0^+$ ($J^P = 1^+$) pole will give rise to a term $s^{-2}$ ($s^{-1}$) in the real part of the amplitude $A_3$ ($A_4$).

In order to derive this result, we express $A_{3,4}$ by means of (83) at high energies in terms of t-channel helicity amplitudes with natural and unnatural parity exchange, respectively,

$$A_3(s,t) \simeq \frac{p}{2(\sin\Theta_t)qs\sqrt{t}}\, T_{(+)} \ , \qquad A_4(s,t) \simeq \frac{1}{2(\sin\Theta_t)s\sqrt{t}}\, T_{(-)} \tag{89}$$

with

$$T_{(\pm)} = -(f^t_{1,-1;\frac{1}{2},-\frac{1}{2}} \mp f^t_{1,-1;-\frac{1}{2},\frac{1}{2}}) \tag{90}$$

where, according to (87) and (85), $T_{(+)}$ contains natural parity exchange and $T_{(-)}$ unnatural parity exchange, respectively.

The partial wave expansions of the even signature part of $T_{(+)}$ (called $T^+_{(+)}$) and of the odd signature part of $T_{(-)}$ (called $T^-_{(-)}$), respectively, look as follows:

$$T^+_{(+)} = \sum_{J=2}^{\infty} (2J+1)d^J_{20}(\cos\Theta_t)\ \frac{1}{2}\ (1+e^{-i\pi J})a^+_{(+)}(J,t) \tag{91a}$$

$$T^-_{(-)} = \sum_{J=2}^{\infty} (2J+1)d^J_{20}(\cos\Theta_t)(1-e^{-i\pi J})a^-_{(-)}(J,t) \ . \tag{91b}$$

Now we assume in accordance with the fixed pole mechanism described in the Appendix the following pole structures of the partial wave amplitudes in the complex J-plane.

$$a^+_{(+)}(J,t) = \sum_i \frac{\beta^+_i(t)\sqrt{J}}{J-\alpha^+_i(t)}\ [\frac{1}{J} + \text{regular terms}] \tag{92a}$$

$$a^-_{(-)}(J,t) = \sum_i \frac{\beta^-_i(t)\sqrt{J-1}}{J-\alpha^-_i(t)}\ [\frac{1}{J-1} + \text{regular terms}] \tag{92b}$$

corresponding to a $J=0$ $(J=1)$ fixed pole in $T^+_{(+)}$ $(T^-_{(-)})$.

The power of the square root in the numerators of (92) have been chosen in agreement with the behavior of the respective Born terms [5].

Insertion of (92) into (91) leads finally by means of standard techniques (Sommerfeld-Watson transformation and Cauchy's residue theorem) to

$$T^{+}_{(+)} = \pi \sum_i (2\alpha^{+}_i(t) + 1)\,\frac{1}{2}\,\frac{1 + e^{-i\pi\alpha_i(t)}}{\sin\pi\alpha_i(t)}\,\gamma^{+}_i(t)\,d_{20}^{\alpha^{+}_i(t)}(-\cos\Theta_t)$$

$$+ \left(\frac{d^{J}_{20}(-\cos\Theta_t)}{\sqrt{J}}\right)_{J=0} \gamma_0(t) + \text{background integral} \qquad (93)$$

with

$$\gamma^{+}_i(t) = \beta^{+}_i(t)\sqrt{\alpha^{+}_i(t)}\left[\frac{1}{\alpha^{+}_i(t)} + \text{regular terms}\right], \qquad \gamma_0(t) = -\sum_i \frac{\beta^{+}_i(t)}{\alpha^{+}_i(t)}$$

and

$$T^{-}_{(-)} = \pi \sum_i (2\alpha^{-}_i(t) + 1)\,\frac{1}{2}\,\frac{1 - e^{-i\pi\alpha^{-}_i(t)}}{\sin\pi\alpha^{-}_i(t)}\,\gamma^{-}_i(t)\,d_{20}^{\alpha^{-}_i(t)}(-\cos\Theta_t)$$

$$+ \left(\frac{d^{J}_{20}(-\cos\Theta_t)}{\sqrt{J-1}}\right)_{J=1} \gamma_1(t) + \text{background integral}$$

with

$$\gamma^{-}_i(t) = \beta^{-}_i(t)\sqrt{\alpha^{-}_i(t) - 1}\left[\frac{1}{\alpha^{-}_i(t) - 1} + \text{regular terms}\right]$$

$$\gamma_1(t) = -\sum_i \frac{\beta^{-}_i(t)}{1 - \alpha^{-}_i(t)} .$$

In agreement with (86) leading Regge trajectories contribute to the even signature part of $A_3$ only. With that and Froissart's bound ($\alpha_i(t) \leq 1$) we obtain by means of (89)-(93) the following high-energy expression for $A_{3,4}$:

$$A_3(\nu,t) = A_{3R}(\nu,t) - \frac{R_0(t)}{\nu^2} + A_3(\nu,t) \qquad (95a)$$

with

$$A_{3R}(\nu,t) = \sum_{\alpha_i(0)\geq 0} b_{3i}(t)\,\frac{1}{2}\left(1 + e^{-i\pi\alpha_i(t)}\right)\nu^{\alpha_i(t)-2}$$

and

$$A_4(\nu,t) = -\frac{R_1(t)}{\nu} + \tilde{A}_4(\nu,t) \tag{95b}$$

where the fixed pole residua $R_i(t)$ and Regge pole residua $b_{3i}(t)$ are real functions of t in the physical s-channel region. The residual terms $\tilde{A}_3$ $(\tilde{A}_4)$ fall off faster than $\nu^{-2}$ $(\nu^{-1})$ for $\nu \to +\infty$.

We will now derive--one by one--the sum rules for $R_0(t)$ and $R_1(t)$. From (95) it follows that both amplitudes $A_{3,4}$ satisfy unsubtracted fixed-t dispersion relations. For the crossing even amplitude $A_3$ we obtain from (49) and (51)

$$A_3(\nu,t) = A_3^{Born} + \frac{2}{\pi}\int_{\nu_0}^{\infty} d\nu' \, \frac{\nu' \, \mathrm{Im}\{A_3(\nu',t)\}}{\nu'^2 - (\nu + i\varepsilon)^2} \; . \tag{96}$$

Insertion of (95) into (96) leads, by means of

$$\frac{2}{\pi} P \int_0^{\infty} d\nu' \, \frac{\nu'^{\alpha_i(t)-1}}{\nu'^2 - \nu^2} = -\mathrm{ctg}\,\frac{\pi\alpha_i(t)}{2}\, \nu^{\alpha_i(t)-2} \quad ,$$

to the relation

$$-R_0(t) + \nu^2\tilde{A}_3(\nu,t) = \nu^2 A_3^{Born} + \frac{2\nu^2}{\pi}\int_{\nu_0}^{\infty} d\nu' \, \frac{\nu' \, \mathrm{Im}\{\tilde{A}_3(\nu',t)\}}{\nu'^2 - (\nu + i\varepsilon)^2}$$

$$- \frac{2\nu^2}{\pi}\int_0^{\nu_0} d\nu' \, \frac{\nu' \, \mathrm{Im}\{A_{3R}(\nu',t)\}}{\nu'^2 - \nu^2} \; . \tag{97}$$

With the explicit form of $A_3^{Born}$ given in (50) we may perform the limit $\nu \to \infty$ in (97) and obtain the desired sum rule

$$-R_0(t) = -\frac{\pi\alpha\kappa^2}{4m^4}\, t - \frac{2\pi\alpha(1+\kappa)}{m^2} - \frac{2}{\pi}\int_{\nu_0}^{\infty} d\nu'\nu' \, \mathrm{Im}\{\tilde{A}_3(\nu',t)\}$$

$$+ \frac{2}{\pi}\int_0^{\nu_0} d\nu'\nu' \, \mathrm{Im}\{A_{3R}(\nu',t)\} \; . \tag{98}$$

Present experimental data do not yet allow the computation of the rhs of this sum rule for arbitrary momentum transfer. But in the forward direction ($t=0$) $\mathrm{Im}\{A_3\}$ is related to the total absorption cross section. By means of (43) and (44) we obtain

$$\mathrm{Im}\{mA_2(\nu,0)+2A_3(\nu,0)\} = \frac{\sigma_{tot}(\nu)}{m\nu} . \tag{99}$$

With our assumption of asymptotic s-channel helicity conservation $A_2$ contains neither any leading Regge trajectory nor a $J=0$ fixed pole, i.e., we have $A_2=\tilde{A}_2$ and instead of (98) a superconvergence relation

$$0 = \frac{2\pi\alpha\kappa}{m^2} - \frac{m}{\pi}\int_{\nu_0}^{\infty} d\nu'\nu' \ \mathrm{Im}\{A_2(\nu',t)\} . \tag{100}$$

If we add (98) and (100) for $t=0$ and use the optical theorem (99) after subtraction of the Regge contribution we get

$$m^2R_0(0) = 2\pi\alpha + \frac{m}{\pi}\int_{\nu_0}^{\infty} d\nu'\tilde{\sigma}_{tot}(\nu') + \frac{m^2}{\pi}\sum_{\alpha_i(0)\geq 0}\frac{b_{3i}(0)\nu_0^{\alpha_i(0)}}{\alpha_i(0)} \tag{101}$$

where $\tilde{\sigma}_{tot}=\sigma_{tot}-\sigma_{tot,R}$.

Controversial statements on the computation of the rhs of this sum rule by means of experimental cross section data and Regge analyses exist. DAMASHEK and GILMAN [10] obtain $m^2R_0 \simeq 2\pi\alpha$ (Thomson limit). With more recent data, TAIT and WHITE [54] performed a similar analysis. They found a high sensitivity of $R_0$ on the $A_2$ intercept (put equal to the P' intercept according to the idea of exchange degeneracy of P' and $A_2$). They obtain, in agreement with ARMSTRONG et al. [41],

$$m^2R_0 \lesssim 2\pi\alpha .$$

MOFFAT and SNELL [55] get $R_0=0$ by including Regge cuts in their analysis. But in agreement with the starting assumption of Kellet we have used above, no one has found an $R_0$ of large magnitude.

If we believe in $m^2R_0(0) = 2\pi\alpha$ for Compton scattering off protons we would expect the corresponding Thomson limit result for other targets, in particular $R_0(0) = 0$ for a neutron target. This latter supposition has been confirmed recently by an analysis of (101) for the neutron case [56]. On the other hand, BIYAJIMA [57] obtains from his analysis the value $R_{0n}(0)/R_{0p}(0) \simeq 2/3$, in agreement with the quark model predictions as will be shown below. The difference between [56] and [57] is the choice of the cutoff energy in the evaluation of the integral in (101). It is interesting that a computation of $R_0(0)$ for a deuteron target seems to deviate from the corresponding Thomson limit but is perhaps consistent with the weak binding limit [58].

Along the same line of thinking we derive a sum rule for the fixed pole contribution to $A_4$. We obtain from (49), (51) and (95b) for the crossing odd amplitude $A_4$ the unsubtracted fixed-t dispersion relation

$$A_4(\nu,t) = A_4^{Born} + \frac{2\nu}{\pi}\int_{\nu_0}^{\infty} d\nu' \, \frac{\mathrm{Im}\{A_4(\nu',t)\}}{\nu'^2 - (\nu + i\varepsilon)^2} \, . \tag{102}$$

By means of the high-energy decomposition (95b) and the same procedure as in the case of $A_3$ we obtain

$$-R_1(t) = \frac{\pi\alpha\kappa^2}{m^3} - \frac{2}{\pi}\int_{\nu_0}^{\infty} d\nu' \, \mathrm{Im}\{A_4(\nu',t)\} \, . \tag{103}$$

The relevant optical theorem has the form

$$\mathrm{Im}\{A_4(\nu,0)\} = \frac{1}{4\nu m}\left[\sigma_{3/2}(\nu) - \sigma_{1/2}(\nu)\right] \, . \tag{104}$$

Finally, our sum rule for $R_1(0)$, expressed in terms of experimentally known quantities, takes the form

$$mR_1(0) = -\frac{\pi\alpha\kappa^2}{m^2} + \frac{1}{2\pi}\int_{\nu_0}^{\infty} d\nu \, \frac{\sigma_{3/2}(\nu) - \sigma_{1/2}(\nu)}{\nu} \, . \tag{105}$$

With $R_1(0) = 0$ we identify (105) as the Drell-Hearn-Gerasimov sum rule (14).

It has already been mentioned in Section 1.1 that the violation of this sum rule for the S-V interference term can be explained in terms of a right signature fixed pole with the quantum numbers of the $A_1$ meson.

One must be careful with the interpretation of nonzero results for the $R_i(0)$ from either (101) or (105) because a moving Regge pole with intercept zero could give the same result.

Theoretical predictions for the fixed pole residua $R_{0,1}(t)$ are strongly model dependent.[12]

First of all we consider the parton model [61] in its simplest form for Compton forward scattering. We assume that the nucleon acts like a gas of free spin-1/2 point-like constituents, called partons. Furthermore it is supposed that the transverse momentum of these partons is small compared to the nucleon momentum in the CMS, i.e., each parton carries a fraction x $(0 \leq x \leq 1)$ of the nucleon four momentum p. From these assumptions we obtain the invariant amplitudes at $t = 0$.

$$A_K(\nu,0) = \sum_N P_N \sum_{i=1}^{N} \int_0^1 dx f_i^N(x) A_K^i(x,\nu,0) \tag{106}$$

where $P_N$ is the probability that there are N partons within the nucleon, $f_i^N(s)$ is the probability of parton i having in an N-parton configuration four momentum xp, and $A_K^i$ is the contribution of the ith parton.

Because partons are point-like and free, the $A_K^i$ are Born-term contributions for spin 1/2-particles with momentum xp. In particular, if we denote by $Q_i$ and $\kappa_i$ charge (in units of e) and anomalous magnetic moment of the ith parton, we obtain the $A_K^i$ from the expressions in Table 3 by multiplication with $xQ_i^2$ and the substitutions $\kappa \to \kappa_i$, $m \to xm$. For $K = 3$ and 4 this yields

$$\nu^2 A_3^i(x,\nu,0) = -\frac{2\pi\alpha Q_i^2(1+\kappa_i)}{m^2 x}, \quad \text{and} \quad \nu A_4^i(x,\nu,0) = \frac{\pi\alpha Q_i^2 \kappa_i^2}{m^3 x^2} \tag{107a,b}$$

From (106) and (107a,b) we conclude that in the framework of our model there is no other contribution than the fixed pole. This very unrealistic result is due to our oversimplified parton picture. But in a more realistic parton model, where other contributions like Regge terms are possible, the

---

12 An attempt of ZEE [59] to prove $R_0 \neq 0$ by starting from some commonly accepted assumptions has been criticized by CREUTZ [60].

structure of the fixed pole terms we derived above remains unchanged.[13] In particular we conclude from (107b), (95b) and (105), in agreement with a recent paper by KHARE [63], that in the framework of parton-like models any deviation from the validity of the Drell-Hearn-Gerasimov sum rule ($R_1 \neq 0$) proves the existence of a fixed pole in $A_4$ and thus by (107b) the existence of an anomalous magnetic moment of partons. Furthermore, as mentioned above already, (107a) allows for a nonzero $J=0$ fixed pole contribution in Compton scattering off neutrons. To be more specific, let us assume a single quark model (parton configuration with $N=3$ and $f_i(x)$ independent of i), then we obtain from (106) and (107a) indeed

$$R_{0,n}(0)/R_{0,p}(0) = 2/3 \ .$$

On the other hand, the presence of right-signature fixed poles in Compton scattering is no criterion for a composite structure of the nucleon. In a recent paper HITE and MÜLLER-KIRSTEN [64] obtained a nonzero result for $R_0(0)$ for Compton scattering off elementary scalar hadrons[14] in the dressed-ladder approximation.[15]

Another explanation for the $J=0$ fixed pole has been offered by CLOSE et al. [66]. They describe the total photo-absorption cross section by an infinite sum over two-particle cross sections

$$\sigma_{tot} = \sum_{A,B} \sigma_{\gamma N \to AB}$$

and attribute the fixed pole to the contact terms contained in the gauge invariant Born-term model for the process $\gamma N \to AB$ in the case of final state particles with higher spins.

---

13 Compare [62] (in these papers no anomalous moments $\kappa_i$ have been taken into account).

14 "Elementary" means in this context a non-vanishing wave-function renormalization constant.

15 Compare [65] for a detailed discussion of the dressed-ladder approximation for electromagnetic processes.

# Appendix

In this appendix we will give arguments for the possibility of a singular behavior of the reduced Pomeron residue $\beta_p^0(t)$ at $t=0$ (compare (80)) due to a wrong-signature fixed pole at $J=1$ (compare [50]).

Suppose only two-particle channels $A_i\bar{A}_i$ occur in the t-channel unitarity equation, which for the process $\gamma\gamma \to A_i\bar{A}_i$ in the even-signature case (described by the partial wave amplitude $a_i^{(+)}(J,t)$) takes the form

$$\mathrm{Im}\{a_i^{+}(J,t)\} = \sum_{K(\mathrm{open})} \left[f_{Ki}^{(+)}(J,t)\right]^{*} \rho_K a_K^{(+)}(J,t) \qquad (\mathrm{A}.1)$$

where $\rho_K$ is the usual phase space factor and $f_{Ki}^{(+)}$ is the even-signature partial wave hadronic amplitude. The unitarity equation for $f_{Ki}^{(+)}$ is given by

$$\mathrm{Im}\{f_{ij}^{(+)}(J,t)\} = \sum_{K(\mathrm{open})} \left[f_{Ki}^{(+)}(J,t)\right]^{*} \rho_K f_{Kj}^{(+)}(J,t) \ . \qquad (\mathrm{A}.2)$$

Equation (A.2) may be solved with the usual N/D-Ansatz

$$f_{ij}^{(+)}(J,t) = \sum_{K(\mathrm{open})} [D^{-1}(J,t)]_{iK} N_{Kj}^{(+)}(J,t) \qquad (\mathrm{A}.3)$$

where N has the left-hand cut and D the right-hand cut (rhc) only with

$$D_{ij}(J,t) = \delta_{ij} - \frac{1}{\pi}\int_{\mathrm{rhc}} dt' \ \frac{N_{ij}^{(+)}(J,t')\rho_j(t')}{t' - t - i\varepsilon} \ . \qquad (\mathrm{A}.4)$$

The corresponding solution of the "weak" unitarity equation (A.1) is given by

$$a_i^{(+)}(J,t) = \sum_{K(\mathrm{open})} [D^{-1}(J,t)]_{iK} \eta_K(J,t) \qquad (\mathrm{A}.5)$$

where again $\eta_K$ has the left-hand cut only.

From (A.3) to (A.5) one reads off the mechanism, which allows fixed poles in "weak" amplitudes but forbids them in hadronic amplitudes.

Suppose $N^{(+)}(J,t)$ develops a fixed pole at $J = J_0$ ($J_0$ real)

$$N_{ij}^{(+)}(J,t) \underset{J\to J_0}{\simeq} \frac{n_{ij}^{(+)}(t)}{J - J_0} .$$

By means of (A.3) and (A.4) this fixed pole is turned into a moving pole for $f_{ij}^{(+)}(J,t)$ which appears at such an angular momentum $J = \alpha(t)$ for which

$$\mathrm{Det}\left[(J - J_0)\delta_{ij} - \frac{1}{\pi}\int_{rhc} dt' \frac{n_{ij}^{(+)}(t')\rho_j(t')}{t' - t - i\varepsilon}\right] = 0$$

holds. On the other hand, a fixed pole in $\eta_K(J,t)$ survives for $a_i^{(+)}$ because, in general, fixed poles in N and $\eta$ are uncorrelated. Furthermore, a fixed pole at $J = J_0$ in $\eta$ and a moving pole in $D^{-1}$ with $\alpha(t_0) = J_0$ are coupled to each other multiplicatively in the amplitude $a_i^{(+)}(J,t)$ according to (A.5). In particular we obtain for the Pomeron contribution to $a^{(+)}(J,t)$

$$a^P(J,t) = \frac{\beta_p^{00}(t)\sqrt{J-1}}{J - \alpha_p(t)}\left[\frac{1}{J-1} + \text{regular terms}\right] . \tag{A.6}$$

Insertion of (A.6) into the partial wave sum (78) and computation of that by means of standard techniques (Sommerfeld-Watson transformation and Cauchy's residue theorem) leads to

$$f_{1,-1}^{tp} = \pi d_{20}^{\alpha_p(t)}(-\cos\Theta_t)(2\alpha_p(t)+1)\,\frac{\frac{1}{2}\left(1+e^{-i\pi\alpha_p(t)}\right)}{\sin\pi\alpha_p(t)}\,\beta_p^{00}(t)\sqrt{\alpha_p(t)-1}$$

$$\times\left[\frac{1}{\alpha_p(t)-1} + \text{regular terms}\right] + \text{background integral} . \tag{A.7}$$

We notice that the residue of the fixed pole contribution at $J = 1$ vanishes, because $d_{20}^J \underset{J\to 1}{\sim} \sqrt{J-1}$ and the signature factor $(1+e^{-i\pi J})/\sin\pi J$ is finite at the wrong-signature point $J = 1$. Therefore, the fixed pole does not show up in the high-energy behavior of $f_{1,-1}^{tp}$ but appears only in the form of the term $1/(\alpha_p(t) - 1)$ within the square brackets in (A.7), leading to a single pole of the original reduced residue $\beta_p^0(t)$ at $t = 0$.

## References


1 See, for example, W.K.H. Panofsky, M. Phillips: Classical Electricity and Magnetism (Addison Wesley Publishing Co., Reading, MA 1959)
2 W. Thirring: Phil.Mag. 41, 1193 (1950). For a textbook representation see J. Bjorken and S. Drell: Relativistic Quantum Field Theory, Section 19.13
3 F. Low: Phys.Rev. 96, 1428 (1954); M. Gell-Mann, M.L. Goldberger: Phys.Rev. 96, 1433 (1954)
4 A.M. Baldin: Nuclear Phys. 18, 310 (1960); S.R. Choudhury, D.Z. Freedman: Phys.Rev. 168, 1793 (1968)
5 G.C. Fox, D.Z. Freedman: Phys.Rev. 182, 1628 (1969)
6 P. Baranov, et al.: Phys. Letters 52B, 122 (1974)
7 M. Gell-Mann, M.L. Goldberger, W. Thirring: Phys.Rev. 95, 1612 (1954)
8 Compare, for example, F.J. Gilman: Physics Reports 4C, 97 (1972)
9 G. Wolf: in Proc. 1971 Intern. Symp. Electron and Photon Interactions at High Energies (Cornell University, 1972). Here the relevant experimental references are found
10 M. Damashek, F.J. Gilman: Phys.Rev. D1, 319 (1970)
11 H. Alvensleben, et al.: Phys.Rev.Letters 3o, 328 (1973)
12 S.D. Drell, A.C. Hearn: Phys.Letters 16, 23 (1966)
13 S.B. Gerasimov: Soviet J.Nucl.Phys. 2, 43o (196o)
14 I. Karliner: Phys.Rev. D7, 2717 (1973)
15 H.D. Abarbanel, M.L. Goldberger: Phys.Rev. 165, 1594 (1968)
16 See, for example, R. Becker: Theorie der Elektrizität, Band II (B.G. Teubner Verlagsgesellschaft, Stuttgart 1959)
17 J. Bernabeu, T.E.O. Ericson, C. Ferro Fontan: Phys.Letters 49B, 381 (1974)
18 See, for example, S.L. Adler, R.F. Dashen: Current Algebras (W.A. Benjamin, Inc., New York, Amsterdam 1968)
19 H. Genz: Nuclear Phys. B97, 541 (1975)
20 Compare, for example, J. Bjorken, S. Drell, Ref.2, Sec.16.7. Equation (22) of the text is obtained from Bjorken-Drell as expression (16.81) by taking into account our different state normalization and using renormalized quantities throughout.
21 We do not consider the problems connected with a possible non-covariance of the retarded commutator of two vector currents. See, for example, D.J. Gross, R. Jackiw: Nuclear Physics B14, 268 (1969); P. Stichel: Commun.Math.Phys. 18, 275 (1970)
22 M. Jacob, G.C. Wick: Ann.Phys. 7, 404 (1959)
23 B.H. Kellet: Phys.Rev. D7, 115 (1973)
24 J. Bjorken, S. Drell: Relativistic Quantum Mechanics (McGraw-Hill, Inc., New York 1964)
25 H.F. Jones, M.D. Scadron: Nucl.Phys. B10, 17 (1969)
26 W.A. Bardeen, W.K. Tung: Phys.Rev. 173, 1423 (1968)
27 R.E. Prange: Phys.Rev. 110, 240 (1958)
28 A. Hearn, E. Leader Phys.Rev. 126, 789 (1962)
29 For a review compare A. Martin: Lecture Notes in Physics, Vol.3 (Springer, Berlin, Heidelberg, New York 1969)
30 J. Baacke, T.H. Chang, H. Kleinert: Nuovo Cimento 12A, 21 (1972)
31 A. Browman, et al.: Phys.Rev.Lett. 33, 1400 (1974)
32 A. Browman, et al.: Phys.Rev.Lett. 32, 1067 (1974)
33 Y. Nagashima: Progr.Theor.Phys. 33, 1828 (1965)
34 Proc. P.N. Lebedev Physics Institute 41 (1968)
35 H. Genzel, M. Jung, R. Wedemeyer, H.J. Weyer: Z.Physik (to be published) Preliminary (uncorrected) data, H. Genzel, M. Jung, K.R. Rausch, R. Wedemeyer, H.J. Weyer: Nuovo Cimento Letters 4, 695 (1972)

36 H. Rollnik: in Proc. XIV Intern. Universitätswochen für Kernphysik, Schladming, 1975, ed. by P. Urban (Springer, Wien, New York)
37 G. Bernardini, et al.: Nuovo Cimento 18, 1203 (1960)
38 W. Pfeil, H. Rollnik, S. Stankowski: Nucl.Phys. B73, 166 (1974)
39 J.S. Barton, et al.: Phys.Letters 42B, 297 (1972). Here references for older measurements in this energy range are found
40 M. Deutsch, et al.: Phys.Rev. D8, 3828 (1973)
41 T.A. Armstrong, et al.: Phys.Rev. D5, 1640 (1972)
42 D.O. Caldwell, et al.: Phys.Rev. D7, 1362 (1973)
43 G. Buschhorn, et al.: DESY 71/52 (1971)
44 D. Lüke, P. Söding: Springer Tracts in Modern Physics 59, 39 (1971)
45 Compare D. Schildknecht: Lecture presented at the International School of Subnuclear Physics, Erice, Sicily (1974) (DESY 74/50)
46 P. Stichel, M. Scholz: Nuovo Cimento 34, 1381 (1964)
47 P.V. Collins, et al.: Phys.Letters 44B, 183 (1973)
48 Compare W. Drechsler: Fortschritte der Physik 18, 305 (1970)
49 L. Bertocchi: in Proc. Heidelberg Intern. Conf. Elementary Particles (North Holland Publishing Co., Amsterdam 1968)
50 H.D.I. Abarbanel, F.E. Low, K.J. Muzinich, S. Nussinov, J.H. Schwarz: Phys.Rev. 160, 1329 (1967)
51 Compare C. Michael: Springer Tracts in Modern Physics 55, 174 (1970)
52 J.W. Moffat, A.C.D. Wright: Phys.Rev. D8, 2152 (1973)
53 M.J. Creutz, S.D. Drell, E.A. Paschos: Phys.Rev. 178, 2300 (1969)
54 N.R.S. Tait, J.N.J. White: Nucl.Phys. B43, 27 (1972)
55 J.W. Moffat, V.G. Snell: Phys.Rev. D6, 859 (1972)
56 C.A. Dominguez, J. Gunion, R. Suaya: Phys.Rev. D6, 1404 (1972)
57 M. Biyajima: Phys.Letters 43B, 197 (1973)
58 J. Gunion, R. Suaya: Phys.Rev. D8, 156 (1973)
59 A. Zee: Phys.Rev. D5, 2829 (1972)
60 M. Creutz: Phys.Rev. D7, 1539 (1973)
61 R.P. Feynmann: Phys.Rev.Lett. 23, 1415 (1969)
62 S.J. Brodsky, F.E. Close, J.F. Gunion: Phys.Rev. D5, 1384 (1972); Phys. Rev. D8, 3678 (1973);
R.P. Hughes, H. Osborn: Nucl.Phys. B54, 603 (1973).
Compare also S.Y. Lee: Nucl.Phys. B45, 449 (1972)
63 A. Khare: Progr.Theor.Phys. 53, 1798 (1975)
64 G.E. Hite, H.J.W. Müller-Kirsten: Phys.Rev. D10, 3761 (1974)
65 P. Stichel: Acta Physica Austriaca, Suppl.XIII, 447 (1974)
66 F.E. Close, H.R. Rubinstein, C. Ferro-Fontan: Nucl.Phys. B85, 461 (1975)

# Status of Interference Experiments with Neutral Kaons

E. Paul

## 1. Introduction

Experiments for studying weak interactions have a long history. Up to the early nineteen fifties, experimental information about weak interactions came in practice only from observations on nuclear $\beta$ decays. After that, with the fast progress of elementary particle physics, studies on weak decays of free-moving particles were soon begun. New possibilities were opened which broadened the basis for the investigations significantly. One could now study in parallel three classes of weak interactions, distinguishable from each other by the particles that take part, i.e., the contents of leptons and hadrons. There are besides the semileptonic class with both hadrons and leptons, which is also studied in nuclear physics, the purely leptonic class, represented by the muon decay, and the purely hadronic (nonleptonic) class, opened via several decay modes of hyperons and kaons.

The decays of neutral kaons, observed probably for the first time in cosmic rays thirty years ago [1], contribute to two of the three classes in a unique way. All kaons are mesonic states (so that the baryon number is zero) which carry with respect to the strong interactions strangeness, that is, $S=+1$ and $S=-1$ for particle and antiparticle, respectively. When decaying, the kaons have to give up their strangeness because there are no lighter particles with $S \neq 0$. Hence the decays cannot be strong; they are indeed weak. For the neutral kaons hadronic and semileptonic decay modes compete with each other (we disregard the extremely rare decay modes; they are unimportant for the effects to be discussed in this part of the book). Explicitly the following decay processes have to be distinguished:

1) The hadronic decays into two pions, i.e., into $\pi^+\pi^-$ and $\pi^0\pi^0$, and into three pions corresponding to the final states $\pi^+\pi^-\pi^0$ and $\pi^0\pi^0\pi^0$.

2) The semileptonic decays into $\pi^+\ell^-\nu$ and $\pi^-\ell^+\bar{\nu}$, where $\ell$ denotes an electron or muon.

Since the middle of the nineteen sixties, a third experimental field has become more and more important for studying weak interactions. This is the neutrino scattering on matter, which yields the basis for the measurement of energy dependence of many weak processes. These are mainly semileptonic marked by the neutrino (antineutrino) and the nucleon(s) in the target. Particle interactions underlying the semileptonic kaon decays are not observed in scattering experiments with high rates. The study of their nature remains for the foreseeable future the domain of decay experiments. This must, on the one hand, be understood as an unwanted restriction, but underlines on the other hand a status of some exclusivity.

About twenty years after the first systematic studies on neutral kaon decays, interest in the phenomena thus detected is undiminished. There is an effect of great theoretical importance, observed so far only in some decay modes of the neutral kaons. This is the famous violation of CP invariance which occurs as an asymmetry in the laws of nature with respect to the combined operation of C, the reflection of all (charge like) quantum numbers distinguishing the particle from its antiparticle, and of P, the reflection in space, called parity operation.

The theoretical interest in CP violation is essentially stimulated by its relation to a more comprehensive operation called CPT (where T stands for time reversal). The validity of CPT invariance is expected on very general theoretical grounds. Since first-generation experiments on weak decays, the knowledge that both P invariance and C invariance are separately broken in weak interactions whereas no such proof could be obtained for T, tended to suggest that CPT invariance could follow from the simultaneous invariance for T and CP operation.

This hypothesis was disproved in 1964 by the famous experiment of CHRISTENSEN et al., in which a tiny amount of CP violation was concluded from the existence of $\pi^+\pi^-$-decays in a long-lived neutral kaon beam [2].

After this discovery some very fundamental questions soon arose for which one could hope to find the answer in further experiments. On the one hand one wanted to know if CPT symmetry on that level of accuracy still held. This can only be realized if some T non-invariance also exists which can compensate the effect of CP violation in the combined operation. On the other hand one wanted to know the source of the CP violation, about which the basic experiment had given rise to many tentative speculations.

In answering these questions the experiments on neutral kaon decays are of primary importance because they have provided for many years more and more detailed information about the CP violation phenomenon. The opportunities for extensively studying the CP violation are based on a unique property of the neutral kaons. They behave in a free-moving beam as a coherent mixture of two distinct particles (with different lifetimes and masses). If both components have common decay modes, one can observe interference between the corresponding amplitudes. The CP violating amplitudes can be studied in this way in all decay modes (assuming the decay rate is large enough) so that one determines magnitude and phase relative to the corresponding CP conserving amplitude. This is described explicitly in Section 2.

CP violation is observed so far only in the two pion decay modes of neutral kaons: first in the $\pi^+\pi^-$-mode, where it was discovered, and second in the $\pi^0\pi^0$ mode. In addition a consistent amount of charge asymmetry in the semileptonic decay modes was detected. Naturally, the measurements on the two pion decay modes are directly important for judging the possible explanations of CP violation. At present the results are converging toward a satisfactory level of consistency, and the accuracy obtained can hardly be improved with current experimental techniques. The accuracy is in fact limited mainly by correlations with two further parameters--the lifetime of the short-lived neutral kaon state and the mass difference between the long- and short-lived kaon states. The measurements of lifetime and mass difference are discussed separately in two sections (3 and 4) preceding the discussion of CP violation measurements on the two pion decays (Sec.5). The interference experiments on the three pion decays which yield so far only upper limits for the CP violation are described in Section 6.

The situation is rather interesting for the interference experiments with the semileptonic decays which are described in Section 7. These experiments are stimulated by the question of the validity of an important selection rule, the so-called $\Delta S = \Delta Q$ rule for strangeness changing semileptonic decays. In the case that this rule is violated on some level (which is not proved), it can cause CP violation simultaneously. This rule which defines the change of the strangeness number in weak processes was introduced by FEYNMAN and GELL-MANN [3]. It is today an element of the "classical" theory of weak interactions [4]. In this theory the vector and the axial vector currents transform like charged members of an SU3-octet so that for the hadrons taking part, the quantum numbers connected with the SU3 group are changed in a well-defined manner. For the semileptonic neutral kaon decay, where the hadrons are the kaon on one side and a charged pion on the other side, this means that the hadronic current connecting $K^0$ with $\pi^-$ will exist but that connecting $K^0$ with $\pi^+$ will not. This law, called the $\Delta S = \Delta Q$ rule, is to be tested for vector and axial vector currents separately. The semileptonic neutral kaon decays yield a unique basis for testing for a vector current.

Two other processes are available for testing for axial vector currents; these are, first, semileptonic $K^+$ decays, where $K^+ \to \pi^+\pi^+\pi^- e^- \nu$ is forbidden due to the $\Delta S = \Delta Q$ rule, whereas $K^+ \to \pi^+\pi^- e^+ \bar{\nu}$ is allowed; and, second, semileptonic $\Sigma^+$ decays with the correspondingly forbidden process $\Sigma^+ \to n e^+ \bar{\nu}$. The $K^+$ decays represent a pure axial vector current whereas the $\Sigma^+$ decays can contain both, a vector and an axial vector current [5].

So far, a violation of the $\Delta S = \Delta Q$ rule is not confirmed by any of the experiments. All three decay processes are in agreement with there being no violation. The upper limits for the $\Delta S = \Delta Q$ violating amplitudes have been significantly decreased by recent experiments, in particular for the neutral kaon decays. The interest in improving the results for the semileptonic kaon decays is essential because the "forbidden" amplitude can here, once it is detected, again be determined in magnitude and phase relative to the "allowed" amplitude by measuring an interference term (Sec.2.3), whereas only the magnitude can be obtained from measuring $K^+$ and $\Sigma^+$ decays, respectively.

Looking at the experimental situation of neutral kaon decays in total, one obtains interesting conclusions concerning the questions about CP violation mentioned above. As is explained in Section 8, it is confirmed that the CP violation is correlated to a T violation so that they can compensate each other, and all results are still consistent with CPT invariance. The situation is less clear for the question concerning the nature of the interaction causing the CP violating transitions (Sec.9). A complete answer can probably not be based on the neutral kaon decays only. There is a need for some information from processes other than the neutral kaon decays. The experiments which seem to be most advanced with respect to such information are those in which the electric dipole moment of the neutron is measured.

In this part of the book we try to illuminate the status of the interference experiments with neutral kaon decays, giving equal weights to three different aspects. These are, first, the discussion of the experimental methods;

second, an estimation of the experimental results; and, third, considerations on current interpretations. The possibilities of new, improved experiments are considered, too. The experiments with neutral kaon decays, which are exclusively directed to the measurement of the parameters of the neutral kaon regeneration in matter, are not included.

# 2. Interference Effects in a Beam of Coherent $K_S^0$ and $K_L^0$ and Possibilities of Measuring Them

## 2.1 Weak Decays and $K^0$-$\bar{K}^0$ Mixing

In the phenomenological description of neutral kaon states, we restrict ourselves to those features closely related to the interference effects to be discussed in this part of the book. We refer to a basic formalism which is essentially useful for analyzing the role of the CP symmetry violation in the decay interactions. This formalism is well known (see for example [6-10]). For the compilation here we rely mainly on the articles of LEE and WU [8] and for many details on the textbook of MARSHAK et al. [4].

Our knowledge of the properties of the neutral kaon is based on consideration of its strong interactions on the one hand and its weak interactions on the other. The strong interactions are observed when the kaons are scattered on matter. In the processes observed, the $K^0$ and its antiparticle $\bar{K}^0$, defined by

$$|\bar{K}^0\rangle = CPT|K^0\rangle$$

are clearly distinguished from each other by exactly one quantum number (conserved in strong interactions). This is the strangeness number S with the values $S=+1$ for $K^0$ and $S=-1$ for $\bar{K}^0$. With this distinction the production of $K^0$ and $\bar{K}^0$ can be separated completely. (We do not consider here $K^0\bar{K}^0$ production). For producing a pure $K^0$ beam, one makes use of the scatterings with a $K^+$ beam, where the strangeness $S=+1$ is conserved. An important example is given by the reaction

$$K^+p \to K^0p\pi^+ \ . \tag{2.1}$$

Correspondingly, the $\bar{K}^0$'s can be produced by $K^-$ scatterings, most simply in the reaction

$$K^- p \to \bar{K}^0 n \ . \qquad (2.2)$$

The strangeness is not conserved in weak interactions. The neutral kaon, which as the lightest particle carrying strangeness has only the possibility of decaying into particles without strangeness, does this weakly. The change of the strangeness is $\Delta S = -1$ for $K^0$ and $\Delta S = +1$ for $\bar{K}^0$, as is obvious, for instance, from their common decay mode into the final state $\pi^+\pi^-$.

There is a second experimental observation caused by weak interactions which characterizes the neutral kaon state in a very peculiar manner. One observes that an originally pure strangeness state as generated by the adequate strong interaction process becomes, when developing in time, "contaminated" by the opposite strangeness state. Thus a neutral kaon beam always represents an admixture of both strangeness states (one may prove this by analyzing the secondary scattering processes which are, due to strangeness conservation, clearly distinct for $K^0$ and $\bar{K}^0$). This admixture is due to transitions between $K^0$ and $\bar{K}^0$ which, through the weak interactions, is not prevented by strangeness conservation. Their existence already follows from the fact that there are common decay modes for $K^0$ and $\bar{K}^0$, namely, decays into two pions and into three pions.

A transition can occur in two steps, each with $|\Delta S| = 1$

$$\begin{array}{ccccc} K^0 & \underset{G}{\leftrightarrow} & \left\{\begin{array}{c} 2\pi \\ 3\pi \end{array}\right\} & \underset{G}{\leftrightarrow} & \bar{K}^0 \\ S=1 & & S=0 & & S=-1 \end{array}$$

The letter G stands for the coupling constant of the usual weak interaction. The mixing is the exceptional attribute of the $(K^0, \bar{K}^0)$ system. There is no similar phenomenon known for other elementary particles. The coherence of the $K^0$ and $\bar{K}^0$ particles, both being present in any neutral kaon beam due to the transition processes, yields the basis for the observation of interference effects. Due to such interference effects the experiments with neutral kaons open a new dimension for studying the features of weak interactions and are certainly irreplaceable up to now.

In order to work out the relations defining the interference experiments, we are going to start with an explicit description of neutral kaon behavior. As already mentioned above, both strong and weak interactions have to be taken into account. As the strength of the strong force is greatly different from that of the weak force (the ratio of the relevant coupling constants is about $10^6$), one can base the description on the rules of perturbation theory. Following this line, one considers the Hamiltonian H for the decay processes to be composed of two parts, a dominating part $H_0$ and a small perturbation V, so that

$$H_0 = H_{STRONG} + H_{EL.MAG.} , \quad V = H_{WEAK} , \quad H = H_0 + V . \tag{2.3}$$

The following outline of the theoretical formalism follows closely the representation given by ROLLNIK [134]. From a "known" solution of the unperturbed problem

$$H_0|\phi_i\rangle = E_i|\phi_i\rangle \qquad i = 0,1,\cdots , \tag{2.4}$$

one derives the solution of the perturbed problem

$$H|\psi\rangle = E|\psi\rangle . \tag{2.5}$$

One obtains then in the simplest case

$$E = E_0 + \langle\phi_0|H|\phi_0\rangle + \sum_{n\neq 0} \langle\phi_0|V|\phi_n\rangle \frac{1}{E_0 - E_n} \langle\phi_n|V|\phi_0\rangle + \cdots$$

or written in terms of operators

$$E = \langle\phi_0|H_0 + V + V \frac{1 - P_0}{E_0 - H_0} V + \cdots|\phi_0\rangle . \tag{2.6}$$

$P_0$ is a projection operator onto $\phi$:

$$P_0|\phi_n\rangle = \delta_{n0}|\phi_n\rangle , \quad \delta_{n0} = \text{Kronecker's symbol}.$$

For the neutral kaons the description is somewhat more complicated because of two properties discussed below.

1) The eigenvalue to the unperturbed problem, called $E_0$, is twofold degenerate. We have $K^0$ and $\bar{K}^0$ as simultaneous eigenstates of $H_0$ and of the strangeness operator S with the eigenvalues

$$E_0^{(1)} = m_{K^0} , \quad S^{(1)} = 1 , \quad E_0^{(2)} = m_{\bar{K}^0} , \quad S^{(2)} = -1 .$$

The masses are equal if $H_0$ is invariant under CPT.[1] Because of the degeneracy in $E_0$, (2.6) has to be replaced by a $(2\times 2)$ matrix with the following components:

$$H_{rs} = \langle\phi_0^{(r)}|H_0 + V + V\frac{1-P_0}{E_0-H_0}V + \cdots|\phi_0^{(s)}\rangle \qquad r = 1,2 \quad s = 1,2 \tag{2.7}$$

with

$$|\phi_0^{(1)}\rangle \equiv |K^0\rangle \ , \qquad |\phi_0^{(2)}\rangle \equiv |\bar{K}^0\rangle \ . \tag{2.8}$$

According to (2.8) the solutions of (2.5) are searched for only in the two-dimensional space spanned by $K^0$ and $\bar{K}^0$, so that the case of interest to us is given by

$$H|\psi\rangle = E|\psi\rangle \tag{2.9}$$

$$|\psi\rangle = a|K^0\rangle + b|\bar{K}^0\rangle \ . \tag{2.10}$$

2) The neutral kaon is not stable. It decays into continuous states like the two-pion state, leading to finite mass widths $\lambda$. One can calculate the masses $m_L, m_S$ and the widths $\lambda_L, \lambda_S$ of the two neutral kaon states in a concise way by diagonalizing the following energy or mass matrix:

$$H_{rs} = \langle\phi_0^{(r)}|R(E_0 + i\varepsilon)|\phi_0^{(s)}\rangle \ , \tag{2.11}$$

with

$$R(E) = H_0 + V + V\frac{1-P_0}{E-H_0}V + \cdots \ ,$$

where $\varepsilon\to 0$ is understood. In (2.11) $E_0$ denotes the zeroth order mass due only to $H_0$. The infinitesimal imaginary quantity $i\varepsilon$ causes a non-hermiticity of $H_{rs}$ and leads to finite (!) imaginary parts of the eigenvalues of (2.5) corresponding to $(1/2)\lambda_L$ and $(1/2)\lambda_S$.

---

[1] The CPT theorem is valid for the neutral kaon with an accuracy of at least $10^{-14}$. One gets this number from the measured mass difference between $K_S^0$ and $K_L^0$ (see definitions later) via the relation $|m_{K^0} - m_{\bar{K}^0}| \simeq |(m_L - m_S)/m_{K^0}|$.

From (2.11) one can obtain important statements about the symmetry behavior of $H_{rs}$, which we will connect to experimental observation for the following three cases: CPT invariance, T invariance and CP invariance.

*Case 1:* CPT invariance

The following equation holds:[2]

$$(CPT)^{-1}R(E)CPT = R(E^*) = R^{\dagger}(E) \qquad (2.12)$$

where * means complex conjugate, † hermetic conjugate. With the definition

$$\bar{K}^0 = CPT\ K^0 \ ,$$

one gets from (2.11) by substitution

$$\begin{aligned} H_{22} &= \langle\bar{K}^0|R(E)|\bar{K}^0\rangle \\ &= \langle K^0|(CPT)^{-1}R(E)CPT|K^0\rangle^* \\ &= \langle K^0|R(E)|K^0\rangle^* \\ &= \langle K^0|R(E)|K^0\rangle \\ &= H_{11} \ . \end{aligned} \qquad (2.13a)$$

The CPT operation does not connect the elements $H_{12}$ and $H_{21}$, since from (2.11) one obtains the empty statements

$$H_{12} = H_{12} \quad \text{and} \quad H_{21} = H_{21} \ .$$

---

[2] The first equality sign in (2.12) follows from the behavior of the single sum terms of R(E): $(CPT)^{-1}\ H_0\ CPT = H_0$; $(CPT)^{-1}\ V\ CPT = V$;

$$(CPT)^{-1}\ V\ \frac{1-P_0}{E-H_0}\ V\ CPT = (CPT)^{-1}\ V\ CPT\ (CPT)^{-1}\ \frac{1-P_0}{E-H_0}\ CPT\ (CPT)^{-1}\ V\ CPT$$

$$= V\ \frac{1-P_0}{E^*-H_0}\ V$$

The second equality follows from the hermeticity of $H_0$ and V.

*Case 2:* T invariance

With $TR(E)T^{-1} = R^{\dagger}(E)$ and

$$T|K^0\rangle = \eta_T|K^0\rangle \; ; \quad T|\bar{K}^0\rangle = \bar{\eta}_T|\bar{K}^0\rangle \; ; \quad |\eta_T|^2 = |\bar{\eta}_T|^2 = 1$$

one gets immediately from (2.11):

$$H_{12} = \eta_T \bar{\eta}_T^* H_{21} \; .$$

From that follows

$$|H_{12}| = |H_{21}|$$

or with a choice of the free phase between $K^0$ and $\bar{K}^0$ that gives $\eta_T \bar{\eta}_T^* = 1$:

$$H_{12} = H_{21} \; . \tag{2.13b}$$

For the diagonal elements one obtains empty statements.

*Case 3:* CP invariance

With $CPR(E)(CP)^{-1} = R(E)$ and

$$CP|K^0\rangle = -\eta_{CP}|\bar{K}^0\rangle \; ; \quad |\eta_{CP}|^2 = 1$$

one obtains from (2.11):

$$H_{11} = H_{22} \quad \text{and} \quad H_{12} = H_{21} \; . \tag{2.13c}$$

So CP invariance implies CPT and T invariance for the H matrix or the other way round, any violation of CPT or T leads automatically to a violation of CP.

In the following, the symmetry characteristics of the energy matrix H will be expressed in terms of the eigenvalues of (2.9). With the usual procedure one finds first the eigenvalues $E_S$ and $E_L$:

$$E_{S,L} = \frac{H_{11} + H_{22}}{2} \pm \sqrt{\frac{(H_{11} - H_{22})^2}{4} + H_{12} + H_{21}} \; . \tag{2.14}$$

The eigenstates of (2.9) $K_S^0$ and $K_L^0$ can be written

$$|K_S^0\rangle = \frac{1}{\sqrt{2(1+|\varepsilon+\delta|^2)}}\{(1+\varepsilon+\delta)|K^0\rangle + (1-\varepsilon-\delta)|\bar{K}^0\rangle\}$$

$$|K_L^0\rangle = \frac{1}{\sqrt{2(1+|\varepsilon-\delta|^2)}}\{(1+\varepsilon-\delta)|K^0\rangle - (1-\varepsilon+\delta)|\bar{K}^0\rangle\} \;. \qquad (2.15)$$

The parameters $\varepsilon$ and $\delta$ are complex numbers fixed by the matrix elements of H. In addition a certain phase convention is used, which can be found for instance in WU and YANG [6]. For small $\varepsilon$ and $\delta$ one gets (see, e.g., [8]):

$$\varepsilon = \frac{H_{12} - H_{21}}{2(E_L - E_S)} \quad \text{and} \quad \delta = \frac{H_{11} - H_{22}}{2(E_L - E_S)} \;. \qquad (2.16)$$

From (2.16) and (2.13) it is obvious how CPT and T invariance are related to $\varepsilon$ and $\delta$, respectively. CPT is directly connected with $\delta$: CPT invariance leads to $\delta = 0$, CPT violation to $\delta \neq 0$. The parameter $\varepsilon$ is not affected by CPT. T invariance leads to $\varepsilon = 0$, violation to $\varepsilon \neq 0$.

For $\varepsilon = \delta = 0$ one gets from (2.15) the new states $K_1^0$ and $K_2^0$:

$$|K_1^0\rangle = (1/\sqrt{2})(|K^0\rangle + |\bar{K}^0\rangle); \quad \text{and} \quad |K_2^0\rangle = (1/\sqrt{2})(|K^0\rangle - |\bar{K}^0\rangle) \;. \qquad (2.17)$$

$K_1^0$ and $K_2^0$ are eigenstates of the CP operator:

$$CP|K_1^0\rangle = |K_1^0\rangle \;, \quad CP|K_2^0\rangle = -|K_2^0\rangle \;.$$

From (2.17) and (2.15) one gets (neglecting terms of higher order in $\varepsilon$ and $\delta$)

$$|K_S^0\rangle = |K_1^0\rangle + (\varepsilon+\delta)|K_2^0\rangle \quad \text{and} \quad |K_L^0\rangle = |K_2^0\rangle + (\varepsilon-\delta)|K_1^0\rangle \;. \qquad (2.18)$$

Equation (2.18) shows the connection with CP violation: $K_S^0$ and $K_L^0$ contain small admixtures (according to $\varepsilon$ and $\delta$) of the "wrong" CP parity. It also shows that the physical states for CP violation are no longer orthogonal like $K_1^0$ and $K_2^0$ because from (2.18) it follows that

$$\langle K_S^0|K_L^0\rangle = 2\,\mathrm{Re}\varepsilon - 2i\,\mathrm{Im}\delta \;. \qquad (2.19)$$

Equation (2.19) connects $\varepsilon$, $\delta$ and the experimentally determined decay amplitudes. This will be given in detail in Section 8 in connection with the discussion of the experimental results.

One can get further interesting information from the development in time of the eigenstates $K_S^0$ and $K_L^0$. This is also determined by the H operator:

$$i \frac{d}{dt} \psi(t) = H\psi(t) .$$

The solutions are

$$|K_S^0(t)\rangle = e^{-iE_S t} |K_S^0(0)\rangle \quad \text{and} \quad |K_L^0(t)\rangle = e^{-iE_L t} |K_L^0(0)\rangle . \qquad (2.20)$$

$E_S$ and $E_L$ are defined by (2.14).

The exponential function in (2.20) contains information about the propagation and decay of the particles. To show this, one splits the H operator into real and imaginary parts

$$H = M - i\Gamma/2$$

and inserts into (2.20) the eigenvalues split in the same way:

$$E_S = m_S - i\lambda_S/2 \ ; \quad E_L = m_L - i\lambda_L/2 . \qquad (2.21)$$

In (2.21) m and $\lambda$ are the eigenvalues of the operators M and $\Gamma$, respectively. By definition they are real numbers. Then $m_S$ and $m_L$ are the masses and $\lambda_S$ and $\lambda_L$ the reciprocal lifetimes of $K_S^0$ and $K_L^0$.

It is important to note that the weak interactions responsible for the lifetimes of $K_S$ and $K_L$ also introduce between these two states a mass splitting. This is entirely analogous to the splitting, measured by Lamb and Retherford, that the coupling with the electromagnetic field introduces between the short-lived state $2p_{1/2}$ and the long-lived state $2s_{1/2}$ of the hydrogen atom (an interesting discussion of the analogy was given by ABRAGAM [11]). Equation (2.14) shows that a mass difference between $K_S^0$ and $K_L^0$ originates from the off-diagonal elements of H

$$H_{12} = \langle K^0|R(E)|\bar{K}^0\rangle \quad \text{and} \quad H_{21} = \langle \bar{K}^0|R(E)|K^0\rangle .$$

(The diagonal elements only contribute if CPT does not hold). The off-diagonal elements are non zero, because the second-order term R(E) already contributes

$$V \frac{1 - P_0}{E - H_0} V \ .$$

The first-order terms $H_0$ and V vanish because of strangeness conservation in strong interactions (for $H_0$) and because of the $\Delta S = \pm 1$ rule for weak interactions (for V). The second-order term contributes via intermediate states of two (or three) pions:

$$<K^0|V|2\pi><2\pi| \frac{1 - P_0}{E - H_0} |2\pi><2\pi|V|\bar{K}^0> \ .$$

From (2.20) and (2.21) the exponential decay law for $K_S^0$ and $K_L^0$ follows:

$$<K_S^0(t)|K_S^0(t)> = e^{-\lambda_S t} <K_S^0(0)|K_S^0(0)>, \quad <K_L^0(t)|K_L^0(t)> = e^{-\lambda_L t} <K_L^0(0)|K_L^0(0)> \ .$$

The parameters $\lambda_{S,L}$ are the reciprocal lifetime $1/\tau_{S,L}$ and, as one can see by integration, the overall counting rate.

The development in time of the eigenstates of strong interaction $K^0$ and $\bar{K}^0$ is not exponential. From (2.15) one finds

$$|K^0> = (1/\sqrt{2})(a^+|K_S^0> + b^+|K_L^0>) \ , \qquad (2.22)$$

$$|\bar{K}^0> = (1/\sqrt{2})(a^-|K_S^0> - b^-|K_L^0>) \ . \qquad (2.23)$$

The parameters $a^+$, $b^+$, $a^-$ and $b^-$ are complex numbers with the value 1 if $\varepsilon = \delta = 0$. (The explicit form of these parameters is given in [8]). For a $K^0$ beam (generated, e.g., by reaction (2.1)) the time dependence is also given by (2.22). The kaon state at time t for the case $\varepsilon = \delta = 0$ is

$$|K(t)> = 1/\sqrt{2}(|K_S^0(t)> + |K_L^0(t)>) \ ,$$

or, using (2.20) and (2.21)

$$|K(t)> = (1/\sqrt{2}) \ e^{-(im_S+\lambda_S/2)t} |K_S^0(0)> + e^{-(im_L+\lambda_L/2)t} |K_L^0(0)>) \ .$$

If we insert relation (2.15) (again for $\varepsilon = \delta = 0$) we get

$$|K(t)\rangle = (1/\sqrt{2})\left(e^{-(im_S+\lambda_S/2)t}(|K^0\rangle + |\bar{K}^0\rangle) + e^{-(im_L+\lambda_L/2)t}(|K^0\rangle - |\bar{K}^0\rangle)\right) \tag{2.22a}$$

The $K^0$ and $\bar{K}^0$ intensity as a function of time is

$$|\langle K^0|K(t)\rangle|^2 = \frac{1}{4}\left(e^{-\lambda_S t} + e^{-\lambda_L t} + 2\cos(\Delta m t)\, e^{-(\lambda_S+\lambda_L)t/2}\right)$$

$$|\langle \bar{K}^0|K(t)\rangle|^2 = \frac{1}{4}\left(e^{-\lambda_S t} + e^{-\lambda_L t} - 2\cos(\Delta m t)\, e^{-(\lambda_S+\lambda_L)t/2}\right) . \tag{2.22b}$$

The quantity $\Delta m$ is given by $\Delta m = m_L - m_S$.

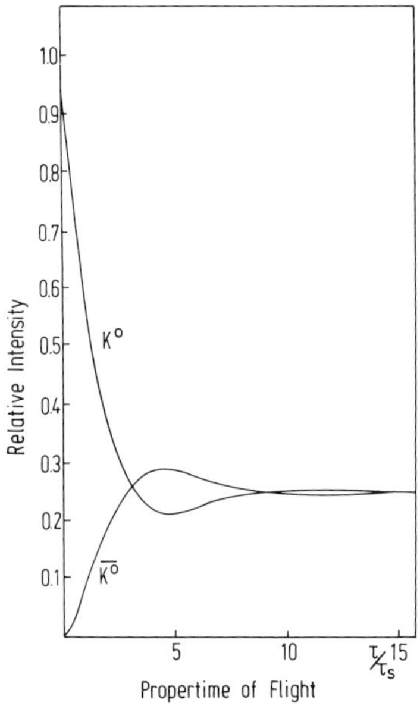


Fig.1 Distributions of $K^0$ and $\bar{K}^0$ intensity as a function of proper time of flight for an originally pure $K^0$ beam

For a $\bar{K}^0$ beam (starting with (2.23)) one gets a corresponding result in which the sign of the interference term is opposite. The time dependence of $K^0$ and $\bar{K}^0$ intensities given in (2.22b) is shown in Fig.1. The interference term causes a so-called "strangeness oscillation", which was not observed for any other elementary particle. Furthermore one can see how the strangeness state originally not present is built up in time. For all experiments subsequently discussed the interference term between $K_S^0$ and $K_L^0$ is of fundamental importance. As one can see from (2.22b) the existence of the mass difference $\Delta m$ causes the periodicity of the time dependence of the interference term. The tiny magnitude of this mass difference in correlation with the lifetimes $\tau_S$ and $\tau_L$ has crucial experimental consequences. The magnitude of the mass difference $\Delta m$ is measured as $5 \cdot 10^{-6}$ eV with a positive sign; that is, the $K_L^0$ is heavier than the $K_S^0$. The magnitude is in agreement with theoretical estimates (see for instance [4]). One can already obtain the correct order of magnitude of $\Delta m$ from the relation between $\Delta m$ and $G^2$:

$$|\Delta m| = \frac{G^2}{4\pi} \sin^2\theta \; m_K^5 \sim 10^{-5} \text{ eV}, \qquad (2.24)$$

with $G = 10^{-5}/m_p$ ($m_p$ = mass of the proton) and $\sin\theta = 0.22$ ($\theta$ = Cabbibo's angle).

This extremely small mass difference is just of the right order of magnitude to allow the observation of the interference effects; if one writes $\Delta m$ in units of the overall counting rate $\lambda_S$ one gets $\Delta m = 0.5\hbar$, so that $\Delta m/\hbar$ and $\tau_S$ are of the same order of magnitude. For this reason the modulation of the amplitude by the term $\cos(\Delta m \cdot t)$ can be observed in the range of a few lifetimes $\tau_S$.

The ratio of the $K_S^0$ and $K_L^0$ lifetimes is two orders of magnitude, the lifetime of the $K_S^0$ being about $10^{-10}$ s and that of the $K_L^0$ about $5 \cdot 10^{-8}$ s. This ratio is roughly expected, taking into account the fact that the main (CP-conserving) decay of the $K_S^0$ in two pions is not important for $K_L^0$ which CP-conserving decays into three pions. The phase space for $2\pi$ is several hundred times larger than that for $3\pi$. The great difference in the lifetimes to a certain extent causes a separation in time and space of the $K_S^0$ and $K_L^0$ decays. How this is utilized in the experiments will be described in the following sections.

## 2.2 Interference in the Pionic Decay Modes Based on CP Violation

An important question is, how to separate the CP violating part of the decay amplitude of $K^0$ and $\bar{K}^0$ into two- and three-pion states, using measurable quantities. One can obtain rather complete information of the magnitude and phase of the CP violating amplitude relative to the CP invariant amplitude by measuring the distribution in time of the relevant decay state. There is always the possibility of starting with a $K^0$ beam (e.g., from reaction (2.1)), or with a $\bar{K}^0$ beam (e.g., from (2.2)). The difference can be expressed by opposite signs (see (2.22) and (2.23)). In the following formulas a double sign is written at each place with opposite signs for $K^0$ and $\bar{K}^0$, and the upper sign will always hold for $K^0$ beam, the lower sign holding for $\bar{K}^0$ beam. If one takes the decay amplitudes of $K^0$ and $\bar{K}^0$ into a certain state f to be

$<f|K^0>$ and $<f|\bar{K}^0>$, one gets the time-dependent decay rate from (2.22) combined with (2.20) and (2.21) as the absolute square of the superposition of $K_S^0$ and $K_L^0$ amplitudes in the following way:

$$R(t) = \left|<f| \begin{matrix} K^0> \\ \bar{K}^0> \end{matrix} \right|^2 = \tfrac{1}{2}\left|a^{\pm}<f|K_S^0(0)>\, e^{-(im_S+\lambda_S/2)t} \pm b^{\pm}<f|K_L^0(0)>\, e^{-(im_L+\lambda_L/2)t}\right|^2$$

$$= \tfrac{1}{2}|a^{\pm}<f|K_S^0(0)>|^2\, e^{-\lambda_S t} + b^{\pm}|<f|K_L^0(0)>|^2\, e^{-\lambda_L t}$$

$$\pm\, a^{\pm}b^{\pm}<f|K_S^0(0)><f|K_L^0>\, e^{-(i\Delta m+(\lambda_S+\lambda_L)/2)t} \quad . \qquad (2.25)$$

The first approximation of $a^{\pm}$ and $b^{\pm}$ is then (taking the definitions of (2.22) and (2.23)): $a^{\pm} = 1 \mp \mathrm{Re}(\varepsilon - \delta)$, $b^{\pm} = 1 \mp \mathrm{Re}(\varepsilon + \delta)$ and $a^{\pm}b^{\pm} = 1 \mp (\mathrm{Re}\varepsilon - i\, \mathrm{Im}\delta)$.

From the decay rates into two pions we know that the amplitudes for $K_S^0$ and $K_L^0$ are very different. The total decay rate of the $K_S^0$ decay is strongly dominated by the amplitudes $<\pi^+\pi^-|K_S^0>$ and $<\pi^0\pi^0|K_S^0>$, whereas the decay amplitudes of the $K_L^0$ into two pions are smaller by a factor of $10^{-3}$ than the corresponding $(K_S^0 \to 2\pi)$ amplitudes. The reason is that the decay of $K_L^0 \to 2\pi$ violates CP invariance.[3] To describe the small $K_L^0$ amplitude relative to the large $K_S^0$ amplitude, one uses the complex ratio of $K_S^0$ and $K_L^0$ amplitudes:[4]

$$\eta_{+-} = \frac{<\pi^+\pi^-|K_L^0>}{<\pi^+\pi^-|K_S^0>} = |\eta_{+-}| \cdot e^{i\phi_{+-}} \qquad \eta_{00} = \frac{<\pi^0\pi^0|K_L^0>}{<\pi^0\pi^0|K_S^0>} = |\eta_{00}| \cdot e^{i\phi_{00}} \, . \qquad (2.26)$$

The decay amplitudes contain two isospin states of the two decay pions, $I = 0$ and $I = 2$ ($I = 1$ would lead to an antisymmetric overall wave function and is therefore excluded). The definitions

[3] To be more precise, $K_L^0$ approximated by $K_L^0 = K_2^0$ has the CP eigenvalue of -1, whereas the decay states $\pi^+\pi^-$ and $\pi^0\pi^0$ have $CP = (-1)^l$ where $l$ is angular momentum. Because angular momentum is conserved and the kaon has spin 0, the eigenvalue of CP for the two pions is well defined to be $CP = +1$.

[4] The connection between the parameters introduced at this point and the parameters $\varepsilon$ and $\delta$ defined in Section 2.1 will be given in Section 8.

$$\varepsilon_0 = \frac{\langle I=0|K_L^0\rangle}{\langle I=0|K_S^0\rangle} \; ; \qquad \varepsilon' = \frac{\langle I=2|K_L^0\rangle}{\langle I=0|K_S^0\rangle} \; ; \qquad \omega = \frac{1}{2}\,\frac{\langle I=2|K_S^0\rangle}{\langle I=0|K_S^0\rangle}$$

and

$$|\pi^+\pi^-\rangle = \sqrt{2/3}\,|0\rangle + \sqrt{1/3}\,|2\rangle \; ; \qquad |\pi^0\pi^0\rangle = \sqrt{1/3}\,|0\rangle - \sqrt{2/3}\,|2\rangle \; ,$$

together with (2.26) lead to

$$\eta_{+-} = (\varepsilon_0 + \varepsilon')/(1+\omega) \; , \qquad \eta_{00} = (\varepsilon_0 - 2\varepsilon')/(1-2\omega) \; . \qquad (2.27)$$

Equation (2.27) shows that $\eta_{+-}$ and $\eta_{00}$ are built up differently, quantitatively to such a degree that $\omega$ is not equal to zero, or else that $\varepsilon'$ contributes relatively to $\varepsilon_0$.

From (2.25) using the definitions (2.26) one obtains the time-dependent decay rate, which can be used to determine $\eta_{+-}$ ($\eta_{00}$):

$$R_{2\pi}(t,\eta) = \frac{1}{2} N_0 \cdot \Gamma(K_S^0 \to 2\pi) \cdot \left\{ e^{-\lambda_S t} + |\eta|^2 \cdot e^{-\lambda_L t} + 2|\eta| \cdot \cos(\Delta m t - \phi) \cdot e^{-(\lambda_S+\lambda_L)t/2} \right\} \qquad (2.28)$$

with $\Gamma$, $\eta$ and $\phi$ either for the $\pi^+\pi^-$ or the $\pi^0\pi^0$ decay. The parameter $N_0$ gives the number of $K^0$ ($\bar{K}^0$) produced at time $t = 0$. The parameter $\Gamma$ stands for the partial decay rate. Different signs of the interference term have to be taken for $K^0$ and $\bar{K}^0$ beam, respectively, as described above. The coefficients $a^\pm$ and $b^\pm$ have been set equal to one for the determination of $\eta$ in lower order. From (2.28) one can see that the interference term is present because of the existence of the CP violating amplitude, expressed by $|\eta|$ and $\phi$. One should note that this causality, as well as the sign difference between $K^0$ and $\bar{K}^0$ beam, does not depend on the decay law being an exponential law, i.e., one can detect the CP violation in general by comparison of the decay rates of a $K^0$ and a $\bar{K}^0$ beam. Fig.2 shows such a comparison of measured $K^0$ and $\bar{K}^0$ decay time distributions [13]. In the experiment considered $K^0$ and $\bar{K}^0$ were produced by $K^+$ ($K^-$) scatterings on carbon in two adjacent runs. Thus the $\pi^+\pi^-$ decays were measured with the same apparatus. The difference between the two decay rates is clearly visible in Fig.2; furthermore, one can see that the expected crossing points occur for $\cos(\Delta m t - \phi) = 0$ with $\phi \simeq \pi/8$.

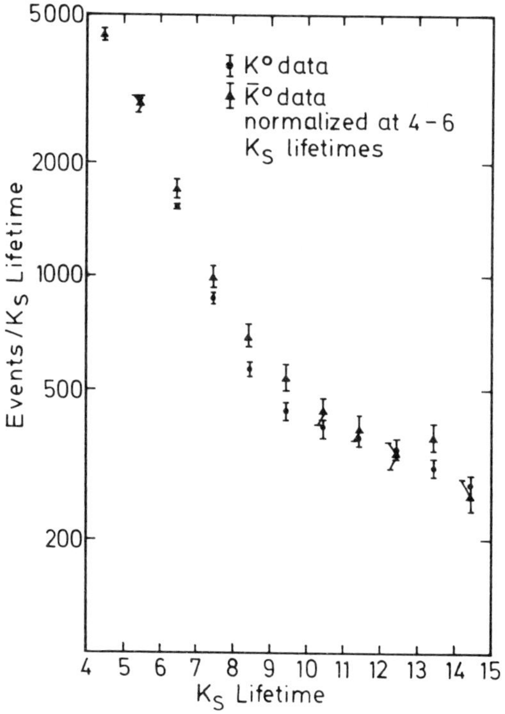


Fig.2 Comparison of $\pi^+\pi^-$ decay rates for originally $K^0$ and $\bar{K}^0$ beam (taken from [13])

To determine $\eta_{+-}$ and $\eta_{00}$ one measures the time distribution of the decays in a time interval with the relatively largest interference term, i.e., between 6 and 15 $K_S^0$ lifetimes. For smaller times the $K_S^0$ component dominates and for larger times the $K_S^0$ component becomes too small for the interference term to give a measurable contribution. The size of $(K_L^0 \to 2\pi)$ amplitude can be determined even for large time intervals (>100 $K_S^0$ lifetimes) by measuring the $(K_L^0 \to 2\pi)$ intensity relative to other $K_L^0$ decays, as it was done in the famous experiment of CHRISTENSEN et al. [2], which led to the detection of CP violation.

The expression in (2.28) demonstrates the strong dependence of the parameters $|\eta|$ and $\phi$ on three other measurable quantities of the $(K^0, \bar{K}^0)$ system, namely the mass difference $\Delta m$ and the two lifetimes $\tau_S$ and $\tau_L$. The determination of the phase $\phi$ is very closely connected to a precise knowledge of the mass difference $\Delta m$. The term which contains $\Delta m$ grows linearly with time. For most experiments which determine $\eta_{+-}$ a variation of the mass difference by one standard deviation results in a change of the phase $\eta_{+-}$ by about 2° (see Sec.5). The influence of an inaccuracy in the $K_S^0$ lifetime $\tau_S$ is twofold;

first, via the exponential function as can be seen in (2.28), and second, indirectly via $\Delta m$, because $\Delta m$ generally has not been determined independently of $\tau_S$. The dependence on $\tau_L$ is generally of minor importance, because in the experiments measurements are made only in time intervals short compared to $\tau_L$. The consequences of a significant change in the $\tau_S$ value which one can accept are of essential importance for estimating the quantitative status of the CP puzzle. In the following sections this will be discussed in more detail.

For the decays into three pions the roles of $K_L^0$ and $K_S^0$ with respect to CP violation are interchanged with respect to the decays into two pions. Now the $K_L^0$ decay is invariant under CP, because the three-pion state exists with CP = -1, whereas the $K_S$ decay violates CP invariance, if the decay leads to the (CP = -1) state of the three-pion system. Such CP violating decays have not been detected up to now. The detection is indeed much more difficult compared to the two-pion channels, because the $(K_S^0 \to \pi^+\pi^-\pi^0)$ decay is possible not only via a CP violating, but also via a CP invariant transition. The CP transformation of three-pion states can be written:

$$CP|\pi^+\pi^-\pi^0> = (-1)^{l+1}|\pi^+\pi^-\pi^0> , \quad CP|\pi^0\pi^0\pi^0> = -|\pi^0\pi^0\pi^0> . \qquad (2.29a,b)$$

The parameter $l$ stands for the relative angular momentum between $\pi^+$ and $\pi^-$, which has to be accompanied by an angular momentum between $\pi^0$ and the $\pi^+\pi^-$ system of equal size and opposite direction.[5] (The sum of angular momenta has to be equal to the kaon spin, i.e., zero). The CP eigenvalue for the $3\pi^0$ state can only be -1; for the $(\pi^+\pi^-\pi^0)$ state it depends on the "internal" angular momentum of the charged pions. The eigenvalue of CP is also -1 for the $\pi^+\pi^-\pi^0$ state when $l = 0$. As a matter of fact this case is much more probable than the case with higher angular momenta: for the relatively small mass of the kaon (~0.5 GeV/$c^2$) angular momenta with $l \geq 1$ are strongly suppressed by the centrifugal barrier [8].

Theoretical estimates [14] give $10^{-2}$ to $10^{-4}$ for the ratio of the CP invariant amplitudes $(K_S^0 \to \pi^+\pi^-\pi^0)/(K_L^0 \to \pi^+\pi^-\pi^0)$. From this it is clear that $K_S^0 \to \pi^+\pi^-\pi^0$ decays have to be more frequent than this order of magnitude to be interpreted directly as CP violating decays, i.e., with a closer

[5] In (2.29a) $l$ enters via the C operation, which exchanges $\pi^+$ and $\pi^-$. As the $3\pi^0$ state is unchanged by C transformation, there is no $l$ in (2.29b).

investigation of the Dalitz plot of the decay particles. Again one defines the amplitude ratios

$$\eta_{000} = \frac{\langle 3\pi^0 | K_S^0 \rangle}{\langle 3\pi^0 | K_L^0 \rangle}, \quad \eta_{+-0} = \frac{\langle \pi^+\pi^-\pi^0 | K_S^0 \rangle}{\langle \pi^+\pi^-\pi^0 | K_L^0 \rangle}. \tag{2.30}$$

With these definitions one finds from (2.25) the time-dependent decay rate in the form usually used for the three-pion decays:

$$R_{3\pi}(t,\eta) = \frac{1}{2} N_0 \cdot \Gamma(K_L \to 3\pi) \cdot$$

$$\cdot \left[ |\eta|^2 \cdot e^{-\lambda_S t} + e^{-\lambda_L t} \pm 2(\mathrm{Re}\eta \cdot \cos\Delta mt - \mathrm{Im}\eta \cdot \sin\Delta mt) \cdot e^{-(\lambda_S + \lambda_L)t/2} \right]. \tag{2.31}$$

The parameter $N_0$ stands for the number of generated $K^0$ and $\bar{K}^0$, respectively. The sign in front of the interference term is positive for $K^0$ and negative for $\bar{K}^0$ beam.

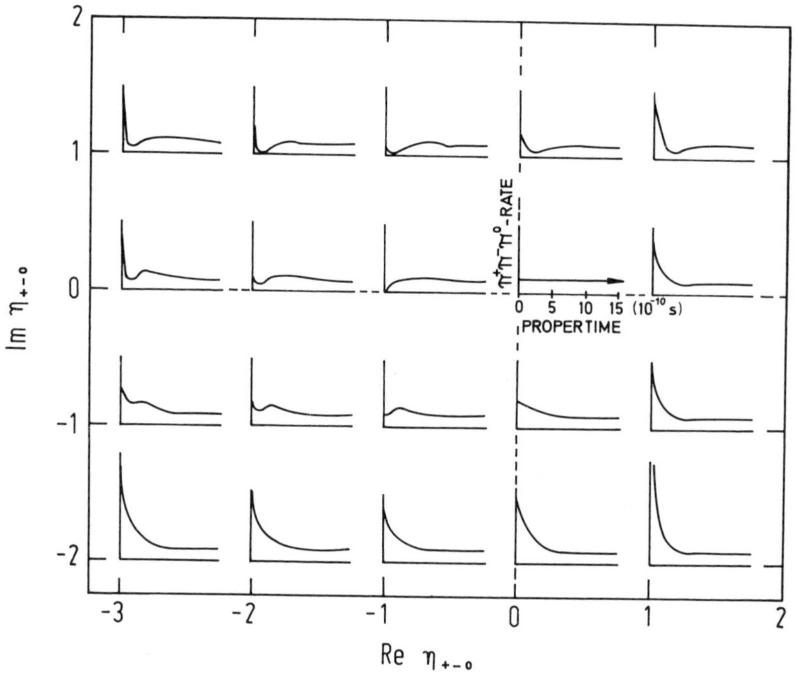


Fig.3 Theoretical $\pi^+\pi^-\pi^0$ decay rates as a function of the proper time of flight for $\eta_{+-0} = 0$ and for various values of $\mathrm{Re}\eta_{+-0}$ and $\mathrm{Im}\eta_{+-0}$ being indicated for the origin points (taken from [19b])

For the quantitative determination of $\eta_{000}$ one measures the time distribution of the three-pion decays for decay times as small as possible, because for small times the $K_S^0$ amplitude relative to the $K_L^0$ amplitude is largest. One can get a more precise idea from Fig.3, in which distributions for different values of $Re\eta_{+-0}/Im\eta_{+-0}$ are shown. These were computed using (2.31) with a positive interference term. The case $Re\eta_{+-0} = Im\eta_{+-0}$ (in the center of Fig.3) shows the nearly flat distribution of the $K_L^0$ decays. The other cases with $\eta_{+-0} \neq 0$ have the common feature that there are structures in the distributions most obvious below 5 $\tau_S$. The essentially sensitive measuring range would be for times below 1 $\tau_S$.

Measurement of these short lifetimes is experimentally very difficult. Therefore the measuring accuracy is not very high. The main difficulties are as follows:

1) The determination of the decay time itself is based on the measurement of $K^0$ momentum and pathlength, and depends mainly on precise measurements of pathlengths not longer than a few centimeters. So far this can only be done, without special expenditure, with instruments like the bubble chamber, which is both target and detector at the same time.

2) For small times the three-pion decays contribute only a very small part compared with the decays into two pions. Therefore the separation of the three-pion decays is made very difficult, and can only be done with sufficient accuracy if one makes considerable experimental expenditure. On top of that the $\pi^+\pi^-\pi^0$ decay (assuming normal measuring accuracy) cannot be easily separated kinematically from the $\pi^+\pi^-\gamma$ decay, which accompanies the strongly populated $\pi^+\pi^-$ channel because of bremsstrahlung effects; furthermore, the $\pi^+\pi^-\pi^0$ decays can easily be confused with the semileptonic $\pi^+\mu^-\nu$ ($\pi^-\mu^+\bar{\nu}$) decays. A proper kinematical separation of the $\pi^+\pi^-\pi^0$ decays is possible if the $K^0$ momentum is determined independently of the decay process. Then the $K^0$ momentum has to be computed from the kinematics of the production process. This type of measurement requires an experimental setup, in which the events can be completely measured, e.g., as in an $H_2$ bubble chamber. For the $\pi^0\pi^0\pi^0$ decay the experimental problems are even greater, because one has to measure at least five of the six $\gamma$'s produced by the $3\pi^0$'s at the $K^0$ decay point. Up to now the only apparatus suitable for that is the heavy-liquid bubble chamber.

## 2.3 Interference in the Semileptonic Decay Modes Based on a Violation of the $\Delta S = \Delta Q$ Rule

In contrast to the pionic decays, one observes the decays into $\pi^- e^+ \bar{\nu}$, $\pi^- \mu^+ \bar{\nu}$, $\pi^+ e^- \nu$ and $\pi^+ \mu^- \nu$, all from the short-lived and the long-lived kaon state. An interesting difference between semileptonic decays with positive lepton charge and those with negative lepton charge shows up when one considers the decays of the strangeness eigenstates $K^0$ and $\bar{K}^0$ in terms of the so-called $\Delta S = \Delta Q$ rule. This rule, postulated by FEYNMAN and GELL-MANN in the context of their famous 4-fermion theory [3], restricts the breaking of strangeness symmetry in semileptonic decays to those processes in which, for the hadron state, the two quantum numbers, strangeness S and charge Q, are changed by the same amount and in the same direction.

For the neutral kaons this has the consequence that $K^0$ and $\bar{K}^0$ cannot decay into the same semileptonic states. According to the $\Delta S = \Delta Q$ rule only the following decays are allowed:

$$K^0 \to \pi^- \ell^+ \bar{\nu} \quad \Delta S = \Delta Q = -1 \ , \qquad \bar{K}^0 \to \pi^+ \ell^- \nu \quad \Delta S = \Delta Q = +1 \ . \tag{2.32}$$

If the $\Delta S = \Delta Q$ rule holds, the following decays are forbidden:

$$K^0 \to \pi^+ \ell^- \nu \quad \Delta S = -1,\ \Delta Q = +1; \quad \Delta S = -\Delta Q$$

$$\bar{K}^0 \to \pi^- \ell^+ \bar{\nu} \quad \Delta S = +1,\ \Delta Q = -1; \quad \Delta S = -\Delta Q \ . \tag{2.33}$$

A comparison between the decays (2.32) and (2.33) shows that the $\Delta S = -\Delta Q$ decays of the $K^0$ lead to the same states as the $\Delta S = \Delta Q$ decays of the $\bar{K}^0$ and vice versa. As $K^0$ and $\bar{K}^0$ are always both present in a kaon beam (see Sec.2.1) it is impossible to identify a single event as $K^0$ or $\bar{K}^0$ decay. One can only measure the $\Delta S = \Delta Q$ transitions via their influence on the time distribution. This method is based on the difference between $K^0$ and $\bar{K}^0$ with respect to their development in time, which was already described in Section 2.1. The state of the neutral kaon at time t can be written according to (2.22a)[6]

$$|K(t)\rangle = \frac{1}{2}\left\{ e^{-(im_S+\lambda_S/2)t}\left(|K^0\rangle + |\bar{K}^0\rangle\right) \pm e^{-(im_L+\lambda_L/2)t}\left(|K^0\rangle - |\bar{K}^0\rangle\right)\right\} \ .$$

[6] One can here put $\varepsilon = \delta = 0$, because this would only result in negligible corrections for the $\Delta S = \Delta Q$ test at the present level of experimental accuracy.

The positive sign is valid for a $K^0$ beam, the negative sign for a $\bar{K}^0$ beam. From this one gets for a certain semileptonic state $\pi\ell\nu$ ($\ell$ = charged lepton) the time-dependent decay rate:

$$N_\ell(t) = |\langle\pi\ell\nu|K(t)\rangle|^2$$

$$= A_\ell \cdot e^{-\lambda_S t} + B_\ell \cdot e^{-\lambda_L t} + [C_\ell \cdot \cos\Delta mt + D_\ell \cdot \sin\Delta mt] \cdot e^{-(\lambda_S+\lambda_L)t/2} \quad (2.34)$$

The coefficients $A_\ell$, $B_\ell$, $C_\ell$ and $D_\ell$ are then expressions containing the amplitudes for $K^0 \to \pi\ell\nu$, $M_\ell$ and for $\bar{K}^0 \to \pi\ell\nu$, $\bar{M}_\ell$:

$$A_\ell = \frac{1}{4}\int\sum_\sigma |M_\ell + \bar{M}_\ell|^2 \cdot \tilde{\varepsilon} \cdot d\rho \ , \qquad B_\ell = \frac{1}{4}\int\sum_\sigma |M_\ell - \bar{M}_\ell|^2 \cdot \tilde{\varepsilon} \cdot d\rho$$

$$C_\ell = -\frac{1}{2}\int\sum_\sigma (|M_\ell|^2 - |\bar{M}_\ell|^2) \cdot \tilde{\varepsilon} \cdot d\rho \ , \quad D_\ell = -\int\sum_\sigma \mathrm{Im}(M_\ell^* \bar{M}_\ell) \cdot \tilde{\varepsilon} \cdot d\rho \ , \quad (2.35)$$

where $d\rho$ stands for the phase space element, $\sigma$ for the spin state and $\tilde{\varepsilon}$ for the detection efficiency ($0 < \tilde{\varepsilon} < 1$). The coefficients can be expressed by the ratio of the ($\Delta S = -\Delta Q$) and ($\Delta S = \Delta Q$) amplitudes.

Starting with the definition

$$x_\ell = \frac{\bar{M}_\ell +}{M_\ell +}$$

one finds for the decays with negative hadron and positive lepton

$$A_\ell = c \cdot |1 + x_\ell|^2 \ , \quad B_\ell = c \cdot |1 - x_\ell|^2 \ ,$$

$$C_\ell = -2 \cdot c(1 - |x_\ell|^2) \ , \quad D_\ell = 4 \cdot c \ \mathrm{Im}(x_\ell) \ . \quad (2.35a)$$

The constant c accounts for the normalization.

One obtains a corresponding result for the opposite charge configuration. With

$$\bar{x}_\ell = \frac{M_\ell -}{\bar{M}_\ell -}$$

one gets

$$A_\ell = \bar{c}|1+\bar{x}_\ell|^2 , \quad B_\ell = \bar{c}|1-\bar{x}_\ell|^2 , \quad C_\ell = 2\bar{c}(1-|\bar{x}_\ell|^2) ,$$

$$D_\ell = 4\bar{c}\ \mathrm{Im}(\bar{x}_\ell) . \tag{2.35b}$$

So to determine the $\Delta S = -\Delta Q$ transitions for a given semileptonic decay state, one has to find the complex number $x_\ell$ or $\bar{x}_\ell$. This can be done via the time-dependent decay rate according to (2.34). If the $\Delta S = \Delta Q$ rule holds, $x_\ell$ and $\bar{x}_\ell$ have to be zero; in that case one can rewrite (2.34)

$$N_{\pi^-\ell^+\nu}(t) = c \cdot \left\{ e^{-\lambda_S t} + e^{-\lambda_L t} \pm \cos(\Delta m t)\, e^{-(\lambda_S+\lambda_L)t/2} \right\}$$

$$N_{\pi^+\ell^-\bar{\nu}}(t) = c' \cdot \left\{ e^{-\lambda_S t} + e^{-\lambda_L t} \mp \cos(\Delta m t)\, e^{-(\lambda_S+\lambda_L)t/2} \right\} . \tag{2.34a}$$

Comparing (2.34a) with (2.22b) one can see that in this case the time distribution of the semileptonic decay rates agrees completely with the intensity distribution for $K^0$ ($\bar{K}^0$), as is to be expected, because the $\Delta S = \Delta Q$ rule according to (2.32) binds the negatively charged state ($\pi^-$) exclusively to the $K^0$ and the positively charged state ($\pi^+$) to the $\bar{K}^0$. The search for a violation of the $\Delta S = \Delta Q$ rule can therefore be performed by measuring the deviation of the experimental distributions of $N_{\pi\ell\nu}(t)$ from (2.34a).

It is necessary to measure both $\pi e\nu$ and $\pi\mu\nu$ decays as, due to the e-$\mu$ mass difference, the parameter $x_e$ does not have the same meaning for an electron as for a muon. This can be seen by connecting $x_\ell$ with the relevant amplitudes. To do this we describe the semileptonic decays in the picture of a current-current interaction of the universal lepton current $L_\lambda$ and a hadronic current $J_\lambda$; in that case the amplitude can be written [5]:

$$M_{\ell^+} = \langle\pi^-|J_\lambda|K^0\rangle\langle\ell^+\nu|L_\lambda|0\rangle , \quad \bar{M}_{\ell^+} = \langle\pi^-|J_\lambda|\bar{K}^0\rangle\langle\ell^+\nu|L_\lambda|0\rangle , \tag{2.36}$$

and the corresponding equations for the decays with negative lepton. For the hadronic part, taking into account the relative parity of kaon and pion in a Lorentz invariant way, one writes

$$\Delta S = \Delta Q: \quad \langle\pi^-|J_\lambda|K^0\rangle = f_+(q^2)\cdot p_\lambda + f_-(q^2)\cdot q_\lambda$$

$$\langle\pi^+|J_\lambda|\bar{K}^0\rangle = \bar{f}_+(q^2)\cdot p_\lambda + \bar{f}_-(q^2)\cdot q_\lambda$$

$$\Delta S = -\Delta Q: \quad \langle\pi^+|J_\lambda|K^0\rangle = g_+(q^2)\cdot p_\lambda + g_-(q^2)\cdot q_\lambda$$

$$\langle\pi^-|J_\lambda|\bar{K}^0\rangle = \bar{g}_+(q^2)\cdot p_\lambda + \bar{g}_-(q^2)\cdot q_\lambda$$

The four-vector $p_\lambda = K_\lambda + \pi_\lambda$ is defined as the sum of the four vectors of kaon and pion, $q_\lambda = K_\lambda - \pi_\lambda$ is the difference. The functions f and g are complex form factors, which can depend on the four-momentum transfer $q^2$ of the kaon to the pion. Possible consequences of such a dependence are discussed by SEGHAL [16].

Following the usual conventions [8,14] one defines the ratios of the form factors:

$$x = \frac{\bar{g}_+}{f_+}\,, \quad y = \frac{\bar{g}_-}{f_+}\,, \quad \xi = \frac{f_-}{f_+} \quad \text{for } \pi^-\ell^+\bar{\nu} \text{ decays}$$

$$\bar{x} = \frac{g_+}{\bar{f}_+}\,, \quad \bar{y} = \frac{g_-}{\bar{f}_+}\,, \quad \bar{\xi} = \frac{\bar{f}_-}{\bar{f}_+} \quad \text{for } \pi^+\ell^-\nu \text{ decays}\,. \tag{2.37}$$

With these definitions one gets in first approximation (see WEBBER [16]):

$$\chi_\ell = x + \frac{m_\ell}{m_K}(y - \xi x)\,, \quad \bar{\chi}_\ell = \bar{x} + \frac{m_\ell}{m_K}(\bar{y} - \bar{\xi}\bar{x})\,. \tag{2.38}$$

From (2.38) one gets for the $\pi e \nu$ decays, neglecting terms with $m_e/m_K$:

$$\chi_e \simeq x \quad \text{and} \quad \bar{\chi}_e \simeq \bar{x}\,.$$

For the $\pi\mu\nu$ decays both terms have to be taken into account, because $m_\mu/m_K \simeq 0.2$ cannot be neglected;

$$\chi_\mu = x + \frac{m_\mu}{m_K}(y - \xi x)\,; \quad \bar{\chi}_\mu = \bar{x} + \frac{m_\mu}{m_K}(\bar{y} - \bar{\xi}\bar{x})\,.$$

This shows that $\chi_e$ and $\chi_\mu$ differ by the term 0.2y. The consequence is that one needs the experiments with $\pi e \nu$ as well as those with $\pi \mu \nu$ decays to check the violation of the $\Delta S = \Delta Q$ rule at two points, i.e., to distinguish between the form factor ratios x and y.

If the semileptonic decays are invariant under CPT, CP or T transformation, one gets unique relations for the $\chi_\ell$ parameters. The experimental check of these relations seems to be important especially because of possible connections with CP violation in two-pion decays, as was emphasized by SACHS [113] (see discussion in Sec.9.1).

From CPT invariance follows

$$M_{\ell^\pm} = \bar{M}_{\ell^\mp}^{*} \quad \text{and} \quad \chi_\ell = \bar{\chi}_\ell^{*} \; . \tag{2.38a}$$

If (2.38a) holds, according to (2.35a) and (2.35b) one measures always the same parameter $\chi_\ell$. Then one can analyze both semileptonic charge states simultaneously, and only half the number of form factors has to be determined. This possibility is used by the experiments with rather small event numbers (nearly all, up to now) for statistical reasons. Of course, by such an analysis the check of CPT invariance is lost; this would certainly be avoided in improved experiments.

From CP invariance it follows that

$$M_{\ell^\pm} = \bar{M}_{\ell^\mp} \quad \text{and} \quad \chi_\ell = \bar{\chi}_\ell \; . \tag{2.38b}$$

If (2.38a) and (2.38b) both hold, the imaginary part of the $\chi$ parameter vanishes

$$\mathrm{Im}\{\chi_\ell\} = \mathrm{Im}\{\bar{\chi}_\ell\} = 0 \; .$$

Invariance under the T operation yields an identical result [8].

We have seen that the $\Delta S = \Delta Q$ rule means that $\chi_e = \chi_\mu = 0$, whereas CP invariance (with CPT invariance) has the weaker consequence that the $\chi_\ell$ parameters must be real. Thus the breakdown of CP invariance can be detected only through $\mathrm{Im}\{\chi_\ell\} \neq 0$ which implies the violation of the $\Delta S = \Delta Q$ rule. On the other hand, the $\Delta S = \Delta Q$ rule can be violated without CP violation. The correlations described so far show clearly the requirements for experiments, which check the $\Delta S = \Delta Q$ rule. These are as follows:

1) Analysis of $\pi e \nu$ *and* $\pi \mu \nu$ decays to get information about the different form factors, which contribute to the amplitudes.

2) Comparison of the two charge states $\pi^+ \ell^- \nu$ and $\pi^- \ell^+ \bar{\nu}$ to test CPT invariance.

3) Measurement of $\chi_\ell$ ($\bar{\chi}_\ell$) as a function of four momentum transfer from kaon to pion and of other Dalitz plot variables.

Experiments which have tried to test the $\Delta S = \Delta Q$ rule are far from being able to fulfill 2) or 3) (for details see Sec.7). The situation is similar to that of three-pion decays, where we saw that it is very difficult to identify the decays under study uniquely and to separate them from "wrong" events.

The real part of $\chi_\ell$ can be determined together with $\mathrm{Re}\varepsilon$ (see Sec.2.1) via the charge asymmetry of the semileptonic decays, defined as

$$\Delta_\ell(t) = \frac{N_{\ell^+}(t) - N_{\ell^-}(t)}{N_{\ell^+}(t) + N_{\ell^-}(t)} . \tag{2.39}$$

The $N_\ell(t)$ are given according to (2.34).

If one develops the time-dependent decay rate into a series around $\varepsilon = 0$ (starting from (2.22) and (2.23), respectively) and neglects the terms quadratic in $\varepsilon$, one obtains the following coefficients:

$$A'_\ell = (1 - \mathrm{Re}\varepsilon)A_\ell , \quad B'_\ell = (1 + \mathrm{Re}\varepsilon)B_\ell , \quad C'_\ell = (1 \pm \mathrm{Re}\varepsilon)C_\ell \quad \text{and}$$

$$D'_\ell = (\pm \mathrm{Im}\varepsilon)D_\ell \tag{2.40}$$

with parameters A to D given by (2.35).

From (2.39) one obtains with (2.34) and (2.40), assuming that $\mathrm{Re}\varepsilon \ll 1$ and $\mathrm{Im}\chi_\ell$ ($\mathrm{Im}\bar{\chi}_\ell$) $\ll 1$, the following approximation (for details see [43]).

$$\Delta_\ell(t) = 2 \frac{1 - |\chi_\ell|^2}{|1 - \chi_\ell|^2} \left\{ \cos(\Delta m t)\, e^{-(\lambda_S + \lambda_L)t/2} + \mathrm{Re}\varepsilon \right\} . \tag{2.41}$$

From that one derives the asymptotic case for very large times $t \geq 20\ \tau_S$:

$$\Delta_\ell^{\mathrm{LIMES}} = 2\,\frac{1-|x_\ell|^2}{|1-x_\ell|^2}\cdot \mathrm{Re}\varepsilon \simeq 2(1+2\mathrm{Re}x_\ell)\mathrm{Re}\varepsilon\ . \tag{2.42}$$

Eq. (2.42) shows that one gets a charge asymmetry if $\mathrm{Re}\varepsilon$ is non zero. If $\mathrm{Re}x_\ell \neq 0$ the charge asymmetry is modified, if this happens in addition to $\mathrm{Re}\varepsilon \neq 0$. As we will see the measurement of the time-dependent charge asymmetry is used to determine not only $\mathrm{Re}\varepsilon$, but also $\mathrm{Re}x_\ell$ and even the mass difference $\Delta m$, when measuring at short decay times ($t \geq 3\ \tau_S$).

## 2.4 Interference with Regenerated $K_S^0$

PAIS and PICCIONI label the production of $K_S^0$ during the transition of a $K_L^0$ beam through matter as regeneration [17]. The reason for the regeneration effect is that the $K_L^0$ as a linear combination of the $K^0$ and $\bar{K}^0$ state according to (2.15) is scattered as $K^0$ or as $\bar{K}^0$. The scattering processes occur more frequently for the $\bar{K}^0$ than for the $K^0$ component according to the difference in total cross section ($\sigma_T(\bar{K}^0N) > \sigma_T(K^0N)$, N = nucleon). The change of ratio of the $K^0$ and $\bar{K}^0$ components is equivalent to building up a new $K_S^0$ component, which again can interfere with the original $K_L^0$ component. By regeneration, one produces conditions similar to $K^0$ or $\bar{K}^0$ beam produced by strong interactions before the dying out of the $K_S^0$ component after a few lifetimes $\tau_S$. According to that, the regeneration technique is also used to measure, for example, the phase of the CP violating amplitude $\phi_{+-}$ in an interference experiment as described in Section 2.2. The regeneration is especially important with respect to the measurement of the mass difference $\Delta m$ between $K_L^0$ and $K_S^0$. This will be described in detail in Section 4. We shall describe the regeneration here only as far as it seems to be necessary for the understanding of the $\Delta m$ experiments. A much wider discussion of regeneration can be found in [8,18].

For a quantitative description of regeneration the eigenstates of the total Hamiltonian have to be found, which in this case has an additional term for the interaction $H_I$ so that

$$H_{\mathrm{total}} = H + H_I\ . \tag{2.43}$$

From (2.43) one finds, if one solves eigenvalue problem and the Schrödinger equation analog to the case without matter (also called vacuum regeneration)

$$|K_S^{0\prime}(t')\rangle = \frac{1}{\sqrt{1+|r|^2}} \{|K_S^0\rangle - r|K_L^0\rangle\} e^{-iE_S' t'}$$

$$|K_L^{0\prime}(t')\rangle = \frac{1}{\sqrt{1+|r|^2}} \{|K_L^0\rangle + r|K_S^0\rangle\} \cdot e^{-iE_L' t'} \quad . \tag{2.44}$$

The parameter r is called the regeneration parameter and can be written [18]

$$r = i \frac{\pi N \Lambda_S}{k} \frac{f(o) - \bar{f}(o)}{1/2 - i\Delta m \lambda_S} \tag{2.44a}$$

where $f(o)$, $\bar{f}(o)$ is the elastic scattering amplitudes for $K^0$, $\bar{K}^0$ in the forward direction; N is the density of the matter; $k = p(\text{kaon})/\hbar$ the wave number of the kaon; and $\Lambda_S = \beta \cdot \gamma \cdot \tau_S$ the average scattering length of the $K_S^0$.

In (2.44) the time t' is measured in the laboratory system. The reason for that is discussed elsewhere (see, e.g., FAISSNER [18]); here it is enough to note that the eigenvalues $E_S'$ and $E_L'$ differ essentially by the factor $1/\gamma$ from the case developed in Section 2.1 for the time in the kaon rest system

$$E_{S,L}' \simeq \frac{1}{\gamma} E_{S,L} \quad . \tag{2.44b}$$

The imaginary parts of the amplitudes $f(o)$ and $\bar{f}(o)$ are connected with the total cross sections via the optical theorem

$$\mathrm{Im}\{f(o)\} = \frac{k}{4\pi} \sigma_T(K^0 N) \quad .$$

If the real parts of $f(o)$ and $\bar{f}(o)$ are small, the difference in the total cross sections already mentioned above can be expressed by

$$|f(o)| < |\bar{f}(o)| \quad \text{and} \quad r \neq 0 \quad .$$

The case relevant to the $\Delta m$ measurements is given by regeneration in a slab of finite thickness. Starting from a pure $K_L^0$ beam going through a slab of thickness L, one finds with (2.44)[7] at time t'

[7] Quadratic terms in r are neglected.

$$|K(t')\rangle = |K_L^{0\prime}(t')\rangle - r|K_S^{0\prime}(t')\rangle = \{|K_L^0\rangle + r|K_S^0\rangle\}\, e^{-iE_L't'} - r|K_S^0\rangle\, e^{-iE_S't'} .$$

Directly behind the slab at time $t' = L/v$ this reads

$$|K(L/v)\rangle = e^{-E_L'L/v}\{|K_L^0\rangle + \rho(L)\cdot|K_S^0\rangle\} \tag{2.45}$$

with the definition

$$\rho(L) = r\cdot\left(1 - e^{-i(E_S'-E_L')L/v}\right) . \tag{2.45a}$$

If one writes (2.45a) using (2.44a) one gets

$$\rho(L) = r\left(1 - e^{(i\Delta m\lambda_S - 1/2)L/\Lambda_S}\right) . \tag{2.45b}$$

The expression (2.45b) shows the development of the $K_S^0$ component with respect to the $K_L^0$ component. At the entrance of the beam into the slab, i.e., for $L = 0$, the component $K_S^0$ is not yet present, as is to be expected.

For the intensity of the $\pi^+\pi^-$ decays of the $K_S^0$-$K_L^0$ mixture produced according to (2.45) at a later time t, which is the kaon time of flight between the end of the regenerator and the decay point (in the kaon rest system) one gets, according to (2.20),

$$R_{\pi^+\pi^-}(t,L) = |\langle\pi^+\pi^-|K(t)\rangle|^2$$

$$\propto \left|\langle\pi^+\pi^-|K_L^0(o)\rangle\cdot e^{-iE_Lt} + \rho(L)\langle\pi^+\pi^-|K_S^0(o)\rangle\cdot e^{-iE_St}\right|^2$$

$$= \left|\eta_{+-}\cdot e^{-iE_Lt} + \rho(L)\, e^{-iE_St}\right|^2 |\langle\pi^+\pi^-|K_S^0\rangle|^2 .$$

When this is combined with (2.26) and the definition

$$\rho(L) = |\rho|\, e^{-i\phi_\rho} ,$$

one gets the result needed for the determination of $\Delta m$:

$$R_{\pi^+\pi^-}(t,L) \propto |\eta_{+-}|^2 e^{-\lambda_L t} + |\rho|^2 \cdot e^{-\lambda_S t} + 2|\eta_{+-}||\rho| \cdot e^{-(\lambda_S+\lambda_L)t/2} \cdot$$

$$\cdot \cos(\Delta m t - \phi_{+-} + \phi_\rho) \,. \tag{2.46}$$

The disadvantage of experiments using the regeneration effect is that they depend also on the regeneration parameters. In general these parameters are not well enough known to neglect their influence on the determination of $\Delta m$. Therefore it is always necessary to measure them as well or to get rid of them by subtraction methods.

# 3. $K_S^0$ Lifetime

## 3.1 Recent Measurements

The lifetime $\tau_S$ of the $K_S^0$ must be determined with high precision in order that its error does not set limits to the accuracy of the other parameters describing the neutral kaon complex. One's knowledge of the CP violation parameters, as observed in the $\pi^+\pi^-$ decay mode, suffers from this correlation; and the present value of the mass difference between $K_L^0$ and $K_S^0$ is also in question due to uncertainties in the measurement of $\tau_S$. The actual problem is that there are new measurements of $\tau_S$ on one hand and earlier measurements on the other which are significantly inconsistent with each other. The new measurements are recent results from two experiments at CERN, one carried through with a bubble chamber, the other with counter techniques. The $\tau_S$ values obtained are in good agreement with each other, but at the same time significantly larger than the averaged value of previous measurements. At first sight the situation looks rather unclear, but a detailed analysis allows the conclusion that the higher $\tau_S$ value is to be favored, as is shown in the following discussion of the most important experiments.

### 3.1.1 The CERN-Oslo-SACLAY Experiment

In 1972 a bubble-chamber experiment for studying the decay time distributions of neutral kaons was completed. This experiment had been carried through with the 2-m chamber at CERN by a collaboration of CERN, Oslo, and SACLAY [19a-e]. The neutral kaons were produced with a low-energy $K^+$ beam in the hydrogen of the chamber. The $K^+$ momentum was set to several values between 1.2 and 1.7 GeV/c. For these energies the simple reaction $K^+p \to K^0 p \pi^+$ is

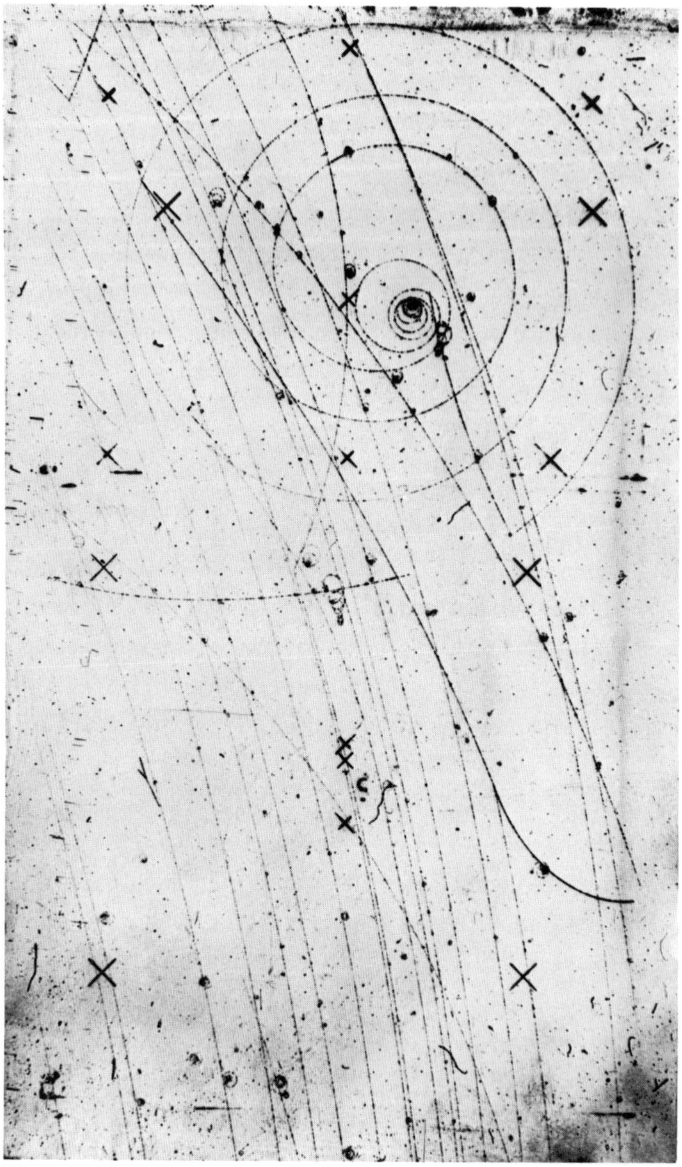

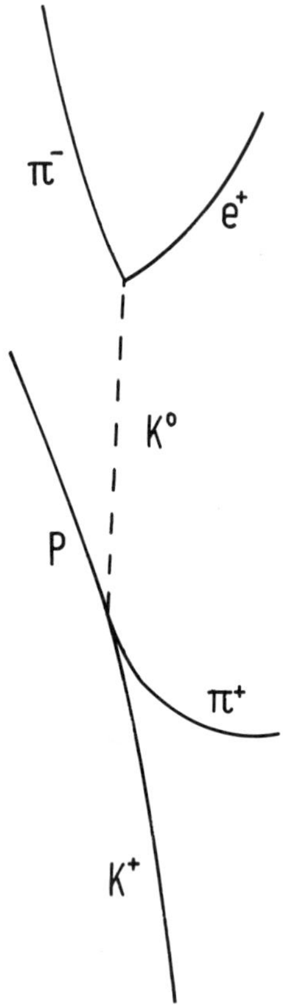


Fig.4

almost the only inelastic process contributing. All pictures showing a candidate for this reaction were measured if a candidate for a kaon decay into at least two charged particles (a so-called $V^0$) was visible on the same frame.

Due to this loose condition, seven decay modes competing with each other are included in the analysis; these are, in practice, first, nonleptonic decays into $\pi^+\pi^-$, $\pi^+\pi^-\gamma$, $\pi^+\pi^-\pi^0$ and, second, semileptonic decays with the final states $\pi^+e^-\nu$, $\pi^-e^+\bar{\nu}$ and $\pi^+\mu^-\nu$, $\pi^-\mu^+\bar{\nu}$, respectively. A total of 70,000 pictures with the production and decay process of the neutral kaon in the same frame was analyzed. Fig.4 shows an example with a semileptonic decay of the kaon state. In order to get a proper distribution of the kaon time of flight it is essential in all cases to calculate this time from the measurement of the distance to the production point and the neutral kaon momentum determined by kinematics either from the production process or from the decay process. It is similarly important to know the acceptance of the $V^0$'s very precisely and, moreover, to avoid misidentification of the decay mode.

The question of how confident one might be in the $\tau_S$ measurement of the COS experiment is considered here in relation to the properties of the device used, that is the 2-m bubble chamber. Several considerations can be made:

1) Measurements of the distance between production and decay point of neutral kaon.

The apparent spatial resolution of the CERN 2-m chamber corresponds to an error of 0.01 cm. This number is so small that in practice another criterion limits the precision, i.e., the mean bubble distance which is of the order of 1 mm. This means that at these low energies, kaons with relatively short times of flight are measurable. With the momentum spectrum shown in Fig.5 a good population is obtained already at eigentimes around 0.1 $\tau_S$. Data with that short time are the best casis for carrying through a precise experiment of $\tau_S$, since the admixture from $K_L^0 \to \pi^+\pi^-$ has least importance here.

2) Determination of the neutral kaon momentum.

The momenta of all charged particles on the one hand and the direction of the $K^0$ and the distance to the decay point on the other are directly calculable from measurements. At the production point the neutral kaon momentum is the only missing quantity so that considering energy and momentum conservation, its determination is overconstraint. It can be obtained in a fitting procedure (three-constraint fit). In a second step the $K^0$ momentum is confirmed by another fit at the decay point with four and one constraint for $\pi^+\pi^-$ and a three-body decay mode, respectively. In this procedure the final

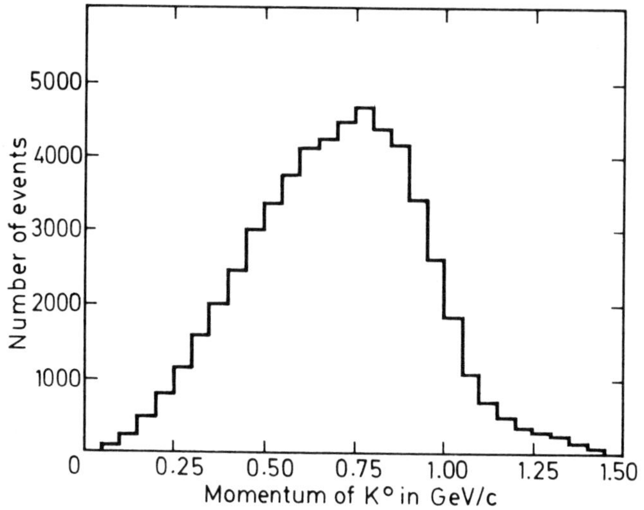


a) Neutral kaon momentum spectrum in the laboratory system (from [19a])

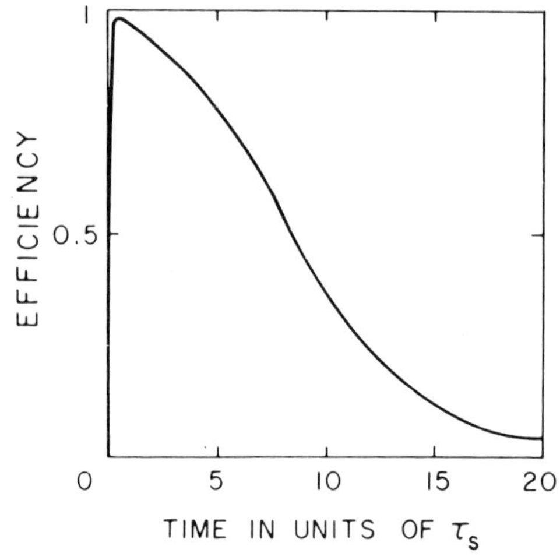


b) Geometrical detection efficiency as a function of $K_S^0$ proper time of flight (from [19c])

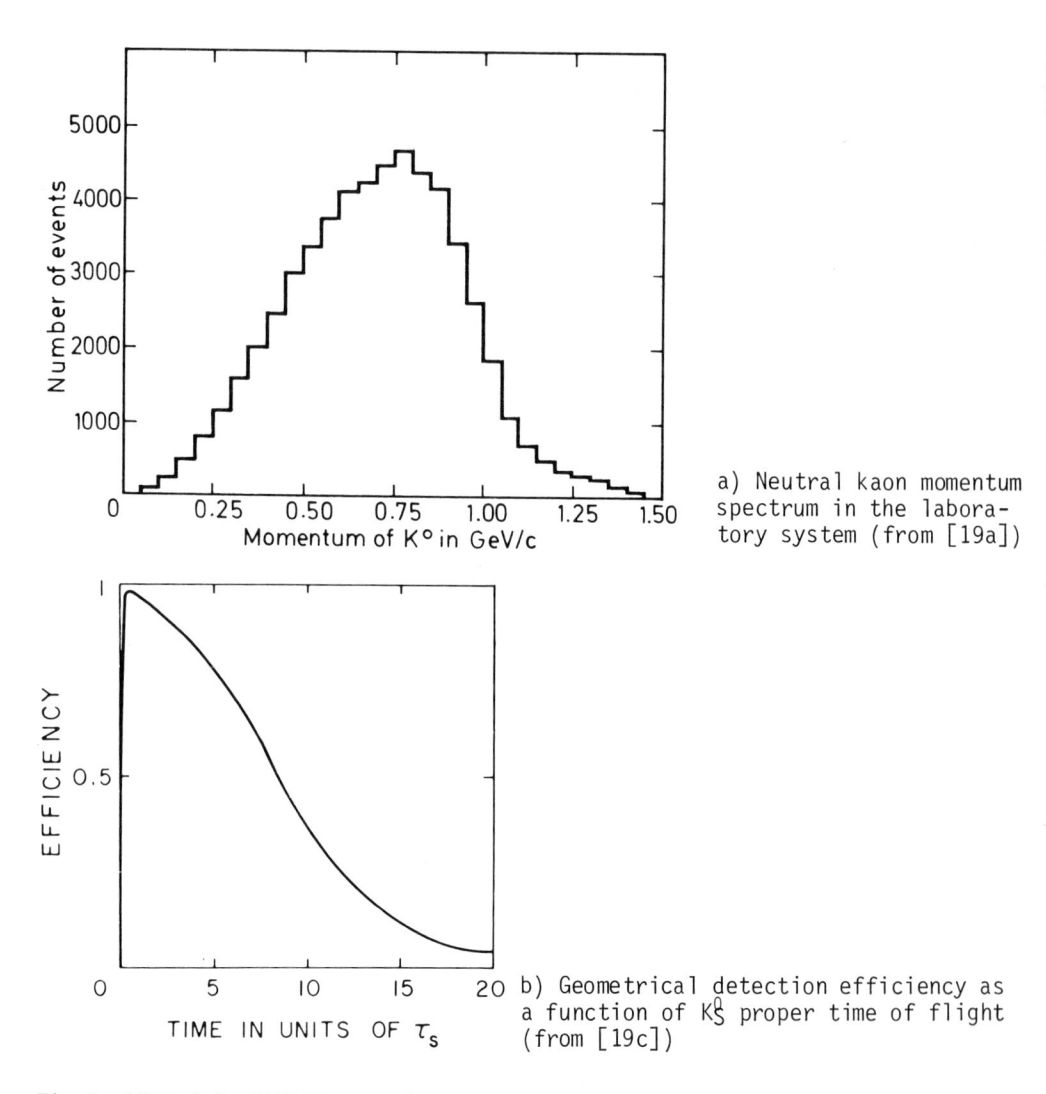

Fig.5 CERN-Oslo-SACLAY experiment

error of the kaon momentum is 5 MeV/c or less. With this precision the probability of relating a $V^0$ to a wrong production point (or the other way around) is estimated to be negligibly small. There was indeed no event remaining in the COS experiment after the kinematics analysis with an ambiguous origin of the $V^0$.

3) Consideration of the $V^0$ acceptance.

The probability for observing a $V^0$ along the line of flight of the neutral kaon is in a good approximation constant for all distances between the production point and the intercept within the limitations of the optical volume of the chamber. From this it follows that for a given event with a definite $K^0$ momentum there exists a time of flight window with the property that the probability to detect the $V^0$ is almost one inside this window and zero otherwise. It is the essential strength of a bubble chamber that the acceptance function is easily determined. In Fig.5b one sees the geometrical detection efficiency as a function of the $K^0$ eigentime which is obtained by adding up the time windows for the single events.

With the momentum spectrum in Fig.5a the averaged decay length is 4 cm for $K_S^0$ and 26 m for $K_L^0$. As can be deduced from Fig.5b one observes from the decays where two charged particles are produced 95% of the $K_S^0$, but only 1.5% of the $K_L^0$. The distribution in Fig.5b shows the expected accumulation of the events at short times of flight; the dip close to zero is caused by a 1-mm cut in the minimum distance between production and decay point, whereas the rapid fall at the larger times of flight is caused by the limitations in the optical volume of the chamber.

4) Completeness of the analysis.

The acceptance problem can be easily disposed if all detectable events are included in the final analysis. It is a well-known difficulty in bubble-chamber experiments to complete the data to better than a few percent. In the COS experiment only about 10 events out of the 70,000 $V^0$'s originally observed were not analysed finally.

5) Separation of the decay modes.

A $V^0$ which was not recognized as background ($e^+e^-$ pair, etc.) was identified with at least one of the four decay modes involved. The mode $\pi^+\pi^-\pi^0$ was separated completely by considering the decay kinematics: knowing the four momenta of neutral kaon, and of the $\pi^+$ and $\pi^-$ at the decay point one calculates the four-momentum of the missing particle, assuming it is a $\pi^0$. The experimental distribution in the missing mass squared is shown in Fig.6 for all three-body decays assigning the pion mass to the charged tracks. The $\pi^+\pi^-\pi^0$ decays show up as a clearly separated signal around 0.02 $GeV^2/c^4$, the mass squared of the $\pi^0$.

The identification of the other four modes was neither so straightforward nor so complete as for $\pi^+\pi^-\pi^0$. In the absence of any $\gamma$ ray detection, the $\pi^+\pi^-$ decay mode forming the basis for the $\tau_S$ measurement cannot be distinguished from the $\pi^+\pi^-\gamma$ mode with small $\gamma$ momentum ($P_\gamma < 30$ MeV/c). But this does not bring about a bias in the $\tau_S$ measurement as long as the $\pi^+\pi^-\gamma$ decays are $\pi^+\pi^-$ decays with subsequent bremsstrahlung. This hypothesis that all

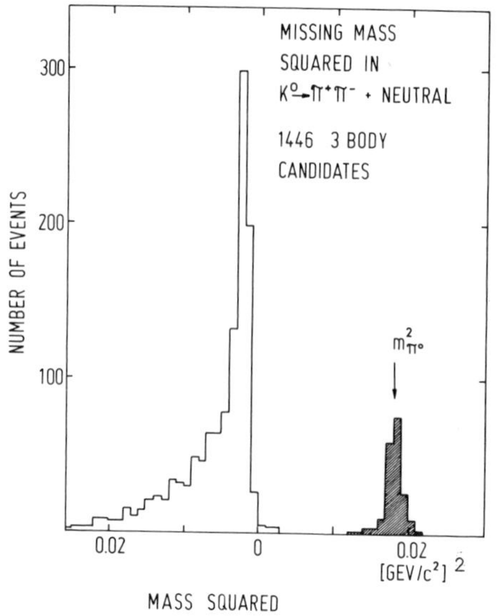


Fig.6 Distribution of missing mass squared for three-body decays of the CERN-Oslo-SACLAY experiment assigning to the charged particles the pion mass. The events with a fit for $K^0 \to \pi^+\pi^-\pi^0$ are drawn black (from [19b])

Table 1 Observed numbers of neutral kaon decays in the COS experiment [19c,e]

| | Decay mode | Total number of events | Fraction of unique identifications |
|---|---|---|---|
| 1 | $\pi^+\pi^-$ | 50,000 | 100% |
| 2 | $\pi^+\pi^-\pi^0$ | 250 | 100% |
| 3 | $\pi^+\pi^-\gamma$ | 300[1] | ≃10% |
| 4 | $\pi^\pm e^\mp \nu$ | 600[1] | 50% |
| 5 | $\pi^\pm \mu^\mp \nu$ | 400[1] | 25% |

[1] These numbers are estimates obtained from the number of $\pi^+\pi^-$ decays via the partial rates [12c]

$\pi^+\pi^-\gamma$ decays are produced in this manner is in good agreement with another result from the COS experiment. It was shown [19d] that the contribution from direct $\gamma$ emission processes to the $\pi^+\pi^-\gamma$ decay mode is less than 5% for $\gamma$ momenta above 50 MeV/c where common models predict a good sensitivity to such a signal. On the contrary, even a small admixture of $\pi\ell\nu$ decays must disturb the analysis of the $\pi^+\pi^-$ decays seriously since the semileptonic decays follow different laws (see Sec.2.3). For the COS experiment this problem is not too important. At short times the rate of the semileptonic decays is still rather small relative to that of $\pi^+\pi^-$ decays (the ratio corresponds to 1.5%).

A summary of the separation of the five decay modes in the COS experiment is given in Table 1. One sees the unique assignment to the decay modes $\pi^+\pi^-$ and $\pi^+\pi^-\pi^0$ as well as the incompleteness in the mutual separation of the three-body modes $\pi^+\pi^-\gamma$ ($P_\gamma > 30$ MeV/c), $\pi e\nu$ and $\pi\mu\nu$ which is hardly avoidable in a bare bubble-chamber experiment. The events with ambiguous decay mode form the complement to the unique sample. They are very useful for several tests for the purpose of confirming the results on the unambiguous events. An important example is described below.

From the discussion under 1 to 5 above, it might be concluded that the COS experiment has successfully optimized the conditions for measuring the $\tau_S$ value with respect to the possibilities defined by the properties of the bare bubble chamber. This is also confirmed by diverse checks for controlling the obvious sources of systematic errors (details are in [19a]). The expression for the $\pi^+\pi^-$ time-dependent rate corresponds to (2.28) with a modification for a small contribution from coherent regeneration in the hydrogen [19a].

$$R_{\pi^+\pi^-}(t,\eta,\rho) = \frac{1}{2} \cdot N_0 \cdot \Gamma(K_S^0 \to \pi^+\pi^-) \cdot$$

$$\cdot \left\{ e^{-\lambda_S t} + \left(|\eta_{+-}|^2 + \frac{4}{9}|\rho|^2\right) \cdot e^{-\lambda_L t} + \left[2\,\mathrm{Re}\left(\eta_{+-} + \frac{2}{3} \cdot \rho\right) \cdot \cos\Delta m t \right.\right.$$

$$\left.\left. - 2\,\mathrm{Im}\left(\eta_{+-} + \frac{2}{3} \cdot \rho\right) \cdot \sin\Delta m t\right] \cdot e^{-(\lambda_S + \lambda_L)t/2} \right\} . \qquad (3.1)$$

In this expression $\rho$ is the so-called regeneration amplitude defined in the usual manner as the ratio of short-lived to long-lived amplitude (see (2.45b)). The factor 2/3 in front of $\rho$ takes into account the fact that only $\pi^+\pi^-$ decays are observed here. To fit $\tau_S$ via the rate given in (3.1), a maximum likelihood method was applied. All additional parameters, i.e., those for describing the contributions from the CP violating ($K_L^0 \to \pi^+\pi^-$) amplitude and from the regeneration in hydrogen as well as $\tau_L$ were set to standard values [12c] without significant loss in accuracy.

The $\pi^+\pi^-$ data and the result of the fit are shown in Fig.7. The value of $\tau_S$ obtained in this fit was corrected upward by 0.3% for the loss of events due to incoherent scattering of $K^0$'s and $\bar{K}^0$'s, respectively. The final value from this experiment was then found to be

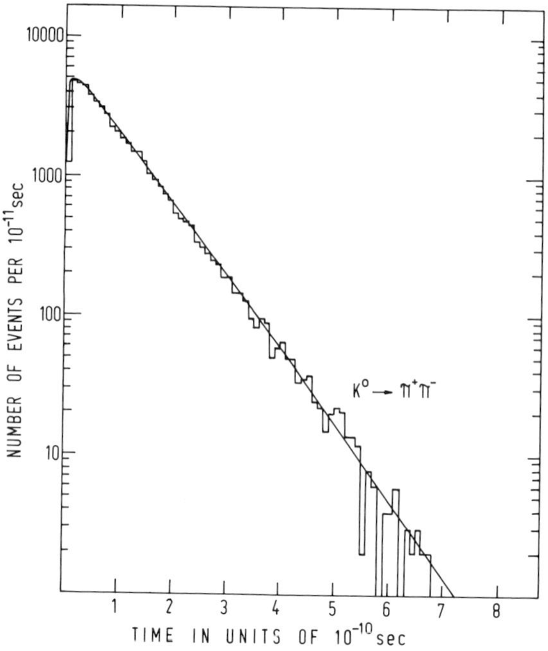


Fig.7 Proper time distribution of the $\pi^+\pi^-$ decays observed in the CERN-Oslo-SACLAY experiment (from [19a]). The curve is the distribution predicted from a maximum likelihood estimate of the mean $K_S^0$ lifetime after correcting for the geometrical detection efficiency shown in Fig.5b

$$\tau_S^{COS} = (0.8958 \pm 0.0045) \times 10^{-10} \text{ s} .$$

The error is purely statistical. The authors claim that systematic errors are negligible. One of the most sensitive tests carried through was the fit of $\tau_S$ using an enlarged data set. Apart from $\pi^+\pi^-$ all the three-body decays consistent with $\pi^+\pi^-\gamma$ or $\pi\ell\nu$ ($\pi\ell\bar{\nu}$) were included. Fig.8 shows the time distribution of the added data and the result of the fit (full line). In this fit $\tau_S$ and the semileptonic rate were determined simultaneously. The numbers quoted are:

$$\tau_S^{COS\ II} = (0.896 \pm 0.005) \times 10^{-10} \text{ s} , \quad \Gamma(K_L^0 \to \pi e\nu + \pi\mu\nu) = (12.4 \pm 0.7) 10^6 \text{ s}^{-1} .$$

The $\tau_S$ value is completely consistent with the result obtained from $\pi^+\pi^-$ decays separately, and the semileptonic rate corresponds to the standard estimate [12c].

Comparing the $\tau_S$ value from the COS experiment with the world average of a all previous measurements which is as quoted in [12a]

$$\overline{\tau_S}^{PREV} = (0.865 \pm 0.0054) \times 10^{-10} \text{ s}$$

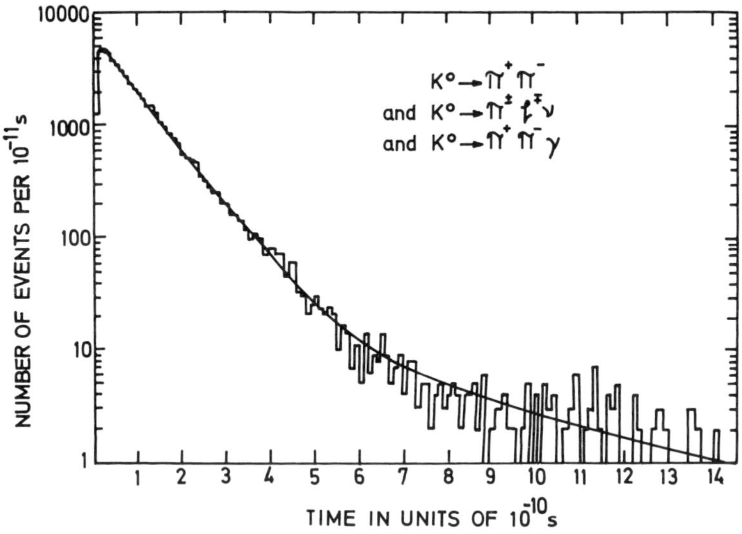


Fig.8 Proper time distribution of all $\pi^+\pi^-$, $\pi^{\pm}\ell\ \nu$ and $\pi^+\pi^-\gamma$ decays observed in the CERN-Oslo-SACLAY experiment (from [19a]). The curve is the predicted distribution based upon a maximum likelihood estimate of the $K_S^0$ lifetime and the partial semileptonic decay rate. The predicted distribution is corrected for the geometrical detection efficiency shown in Fig.5b

one calculates a difference of $0.03 \cdot 10^{-10}$ s corresponding to five standard deviations.

### 3.1.2 The CERN-Heidelberg Experiment

A CERN-Heidelberg collaboration recently completed a large counter experiment at CERN for the study of neutral kaon decays, mainly $\pi^+\pi^-$ and $\pi\ell\nu$ ($\pi\ell\bar{\nu}$). A description of the experiment and the numerous results are given in [20a-e]. One of the parameters determined was the lifetime $\tau_S$ to be discussed here. Several other results from this experiment are discussed in the following sections. The experiment was carried through with a short neutral beam from the CERN proton synchrotron containing neutral kaons in the momentum range from 3 to 15 GeV/c. The main elements of the experimental setup can be seen in Fig.9. The neutral kaons were produced with an external proton beam of 24 GeV/c by scattering the protons on a platinum target (of $4 \times 4$ mm$^2$ cross section and 4.5 cm length). The secondaries were selected in a space angle of 75 mrad by means of a uranium collimator under a magnetic field of 20 KGauss. Downstream to the collimator there was a 9-m long decay volume filled with helium. The tracks of the charged decay products were measured in a spectrometer consisting of proportional wire chambers and a bending magnet.

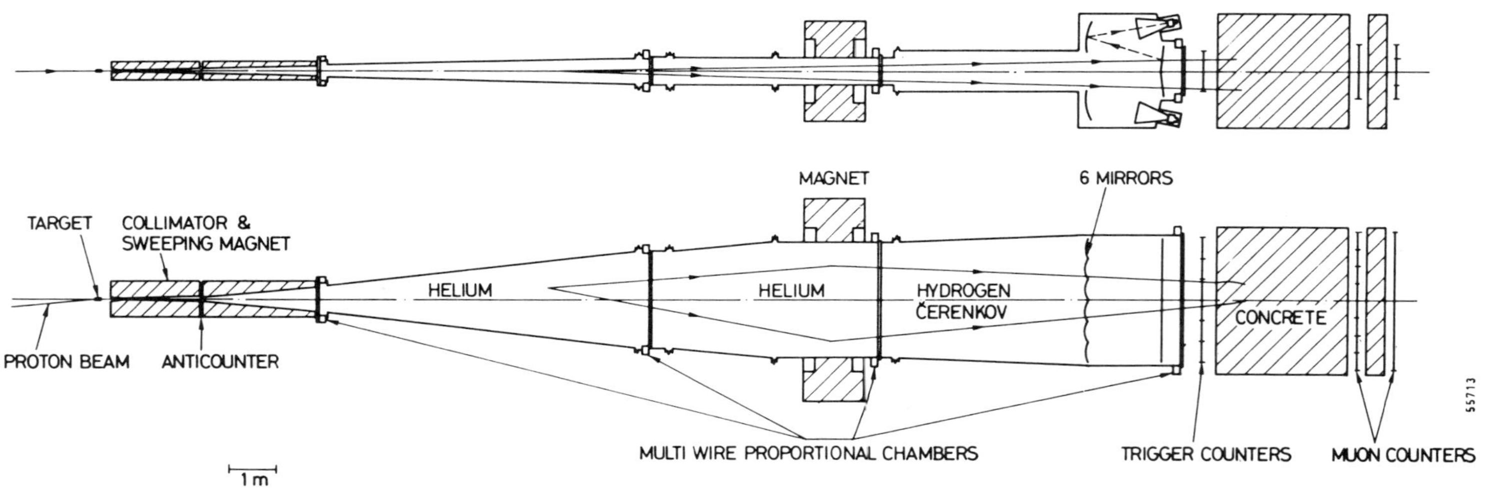


Fig.9 Experimental layout of the CERN-Heidelberg experiment in side view and top view (from [20a])

The spectrometer section was followed by a Cerenkov counter filled with hydrogen at atmospheric pressure for the detection of electrons, and finally, by a muon detector consisting of absorber materials and hodoscopes in front of it and behind.

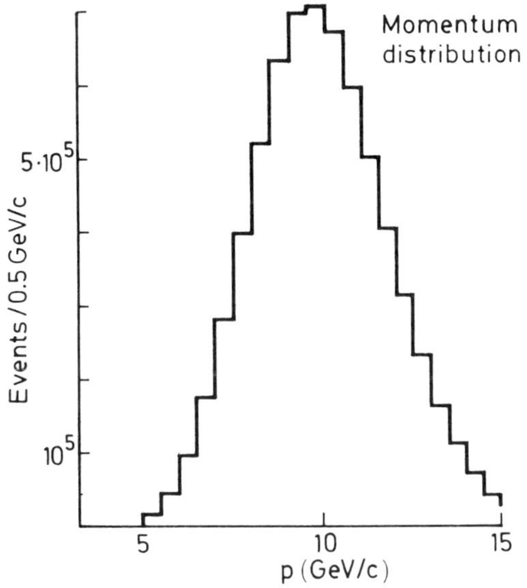


a

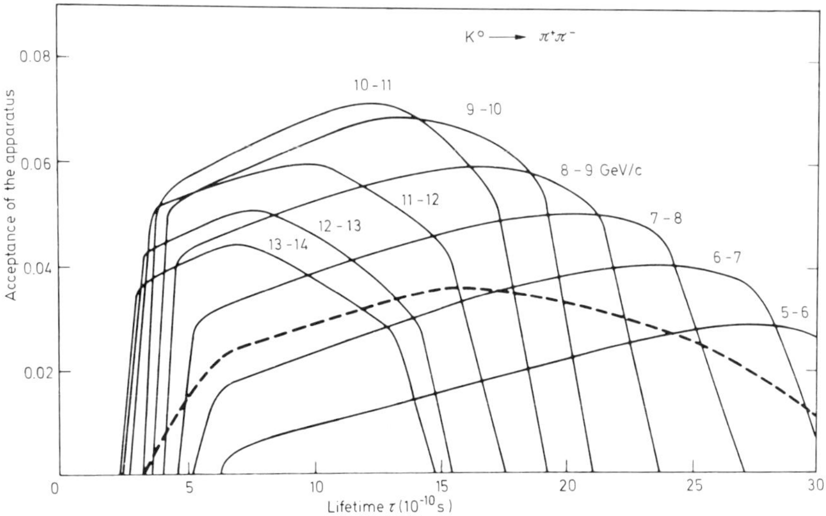


b

Fig.10 CERN-Heidelberg experiment: a) Neutral kaon momentum spectrum in the laboratory system (from [20b]); b) calculated acceptances as a function of the proper time of flight for various neutral kaon momenta (from [20b]). The dashed line represents the average acceptance of the apparatus.

To measure $\pi^+\pi^-$ decays [20b], a decay volume was used which ranged from 2.2 to 11.6 m downstream to the target so that the kaon eigentimes between 4 and 40 $\tau_S$ were populated. A total of $6 \cdot 10^6$ triggers was taken, yielding for each time interval of one $\tau_S$ statistics of at least 4,000 $\pi^+\pi^-$ decays. With respect to the correspondingly small statistical errors the analysis of the data depended now mainly on the precision in determining the acceptance of the apparatus. To get this acceptance, it was necessary to study Monte Carlo-generated events extensively. In particular its dependence on the kaon momentum was relatively strong, as can be seen from the results in Fig.10b. The acceptance reaches its largest value of 0.07 for kaons of 10 GeV/c where, as shown in Fig.10a, the momentum distribution peaks, too.

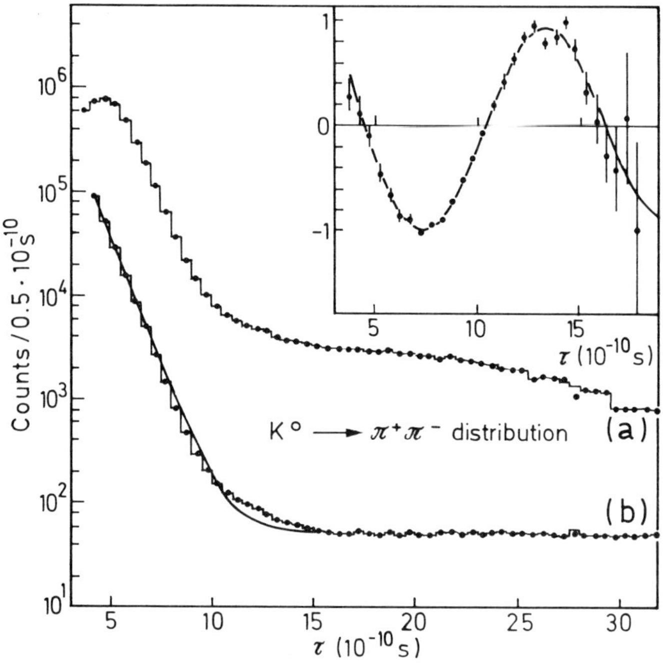


Fig.11 Proper time distribution of the $\pi^+\pi^-$ decays accepted in the CERN-Heidelberg experiment (from [20b]). The events (histogram) and the fitted distribution (dots) are indicated by (a). The events corrected for detection efficiency are indicated by (b) and shown together with the fitted distribution with interference term (dots) and without interference term (solid line). The insert shows the interference term as extracted from data (dots) and fitted term (line).

The time distribution of the $\pi^+\pi^-$ decays with and without the correction for the incomplete acceptance is shown in Fig.11. The points correspond to the best fit obtained. The expression underlying this fit follows from (2.28), if one takes into account that the interference term occurs with both signs, in practice, cancelling each other to the degree that the production cross sections for $K^0$ and $\bar{K}^0$ in the target are of the same magnitude. The coefficient in front of the interference term is given by

$$K(p_K) = \frac{I_{K^0}(P_K) - I_{\bar{K}^0}(p_K)}{I_{K^0}(p_K) + I_{\bar{K}^0}(p_K)}$$

with $p_K$ = kaon momentum and I = intensity at the production point. It is not known a priori and must be determined by the fit.

So the dots drawn in Fig.11 correspond to the result of the fit obtained with the following expression for the rate:

$$R(t,p_K) = \left[I_{K^0}(p_K) + I_{\bar{K}^0}(p_K)\right] \cdot \varepsilon(t,p_K) \cdot \left\{e^{-\lambda_S t} + |\eta_{+-}|^2 \cdot e^{-\lambda_L t} + 2 \cdot K(p_K)|\eta_{+-}| \cdot \cos(\Delta m t - \phi_{+-}) \cdot e^{-(\lambda_S+\lambda_L)t/2}\right\} . \tag{3.2}$$

The lifetime $\tau_S$ was then determined simultaneously with the parameters of the CP violation $\eta_{+-}$ and $\phi_{+-}$ and with $K(p_K)$ yielding the following results:

$$\tau_S^{CH} = (0.894 \pm 0.005) \cdot 10^{-10}\ \mathrm{s}\ , \qquad |\eta_{+-}| = (2.30 \pm 0.035) \cdot 10^{-3}$$

$$\phi_{+-} = (49.4 \pm 1.0)^0 + \left(\frac{\Delta m - 0.540}{0.540}\right) \cdot 305^0 . \tag{3.3}$$

The errors take into account the systematic uncertainties. The value for $\tau_S$ is in excellent agreement with the result obtained in the COS experiment so that it disagrees in the same way with the previous world average (see Subsec. 3.1.1). In the event that the recent $\tau_S$ measurements from the COS experiment and from the CH experiment are correct, the question, what was wrong in the previous experiments, has to be answered. It is discussed in the next section.

## 3.2 Comparison with Previous Measurements

The lifetime of the $K_S^0$, $\tau_S$ was measured many times during the last 15 years, with bubble chambers as well as with sophisticated counter techniques. In

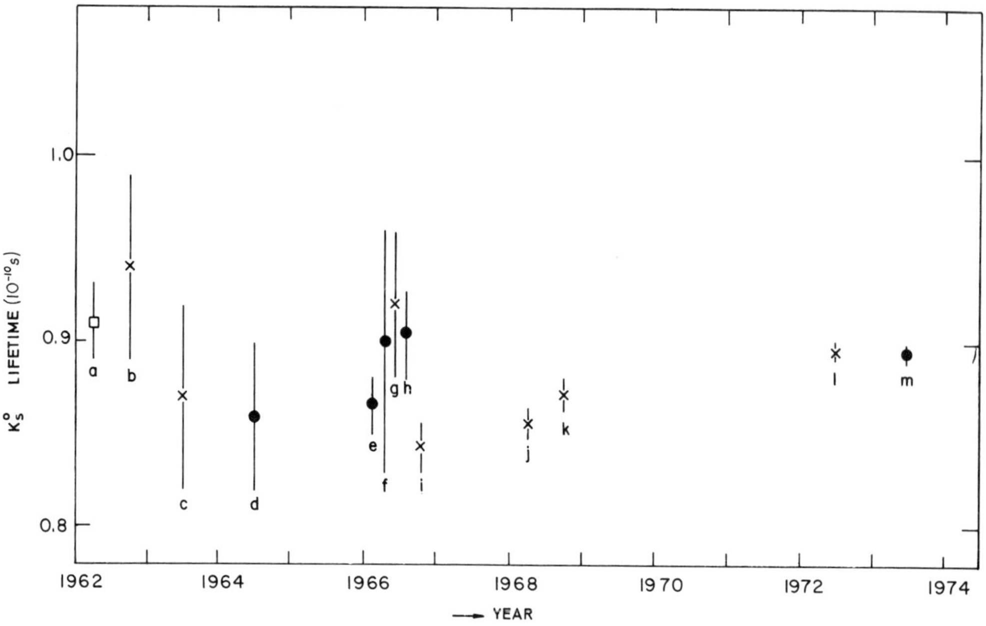


Fig.12 Measurements of $K_S^0$ lifetime $\tau_S$ in chronological order. The technique used is indicated as follows: × means bubble chamber, ● counter setup and □ both techniques. The letters stand instead of reference numbers: a to k for [21-29], l for [20b] and m for [19a]

Fig.12 the measurements are shown collected in a chronological order. Each point with its error bar corresponds to a single experiment except the first point which stands for the average of the experiments finished before the CERN conference in 1962 [21].

When calculating the world average for $\tau_S$ on the basis of the experiments before Kirsch et al., in 1966, one finds the surprisingly large value of

$$\overline{\tau_S}^{66} = (0.890 \pm 0.010) \cdot 10^{-10}\ \mathrm{s}\ .$$

This value is perfectly consistent with the two new results discussed in Section 3.1. Consequently, the shift of $\tau_S$ to the significantly smaller level can only be caused by the more recent measurements. There are only three measurements contributing, but all yielded small $\tau_S$ values which are consistent with each other. The experiments were performed with bubble chambers, two of them with a $\bar{p}$ beam in hydrogen (KIRSCH et al. [29] and DONALD et al. [30]), the third one with a $K^+$ beam in deuterium [31]. Studying the published papers from these experiments one finds for the $\bar{p}p$ experiments possibilities of overlooked systematic effects, whereas such hints are much less obvious for the $K^+$ experiment (which satisfies criteria for the smallness of systematic errors as discussed in Subsec.3.1.1).

The possible bias in the neutral kaon decay time distribution for the $\bar{p}p$ experiments can be discussed in terms of Criteria 2 and 4 defined in Subsection 3.1.1.

Criterion 2 - In $\bar{p}p$ experiments the production processes of neutral kaons are annihilations with complicated final states which do not provide in general the kinematical determination of the three-momentum of a missing neutral kaon. The direction of the kaon and its momentum have to be calculated primarily with the decay process. In this procedure it seems much more likely than in the other case (where the production process yields already the kaon three momenta) that a $V^0$ is connected to the wrong production point. The bias in the decay time distribution originating from wrongly analyzed $V^0$'s cannot easily be estimated, but one cannot exclude the possibility that even a small admixture of such $V^0$'s, say for instance 1%, yields serious distortions.

Criterion 4 - From the analysis procedure of bubble-chamber pictures it is well known that events with a large distance between production and decay point are more easily lost than the events with short kaon path lengths. Such a loss of events can depopulate the longer decay times. There is already some non-negligible probability for such losses during the first step of the analysis, where the pictures are scanned for the desired topologies. A $V^0$ separated from the production process by more than a few cm can be overlooked. Moreover with increasing distance it is more difficult to identify the right production process, in particular when more than one candidate is on the frame. The final decision has sometimes to be based on the measurement of all interactions visible on the frame. In practice it may be necessary to remeasure the same events several times in order to minimize the biases. This was not achieved for the experiment of Kirsch et al., where the determination of $\tau_S$ is based on data measured only once. It was perhaps also not sufficiently achieved for the experiment of Donald et al., supposing that the analysis was carried through at the level of a good experiment for studying strong interactions.

### 3.3 Possibilities for Further Improvements

The error in $\tau_S$ value, measured in the COS experiment, corresponds to 0.5% and is given by the statistical error of the $5 \cdot 10^4$ events included in the analysis. Systematic errors are negligibly small. From detailed studies of the 2 m chamber at CERN [32], one can estimate that this means that one should have systematic errors contributing to the $\tau_S$ error of at most 0.1 or 0.2%. Therefore it seems possible to decrease the present $\tau_S$ error by a factor of two or three via an increase in statistics by another order of magnitude. But the experience with the COS experiment shows that the systematic errors can only be this small if extreme care is taken in order to make a complete analysis of the bubble chamber pictures and to rescue the fraction of misinterpreted events, as it was described in Section 3.1. To satisfy requirements a high statistics experiment (for instance with $10^6$ events) can hardly be carried through in the way the COS experiment was performed, because the conventional procedure for the analysis starting from the scan of the pictures, going then via the measurement with either automatic machines like HPD or manual devices like IEP or SOM for the remeasurements in several iterations is very time- and man-power consuming. But on the other hand, such a high statistics experiment might be possible with a reorganized procedure of analysis as is in principle forseen with new device systems like ERASME [33]. An improvement on the side of counter techniques depends of course not on statistics but on the further reduction of systematic errors. The CH experiments demonstrate that the level of a few tenths of a percent is already extremely difficult to reach. It is not so obvious at present how to do better.

# 4. $(K_L^0 - K_S^0)$ Mass Difference

### 4.1 Methods Used for Determining the Mass Difference

The strong suppression of direct $\Delta S = 2$ transitions (see [6]) leads to an extremely small mass difference $\Delta m = m_{K_L^0} - m_{K_S^0}$. It has a value of $\Delta m = 5 \cdot 10^{-6}$ eV (equal $0.5 \cdot 10^{10}\ \hbar s^{-1}$) which corresponds to $10^{-14}\ m_{K_S^0}$. The tiny mass difference is responsible for the occurrence of an interference term in the time-dependent decay rate resulting from the superposition of $K_S^0$ and $K_L^0$ amplitudes. In order to determine $|\Delta m|$ this interference term has to be measured quantitatively. For $\pi^+\pi^-$ decays most methods are based on interference with a $K_S^0$ amplitude produced by regeneration in matter (a short introduction to this

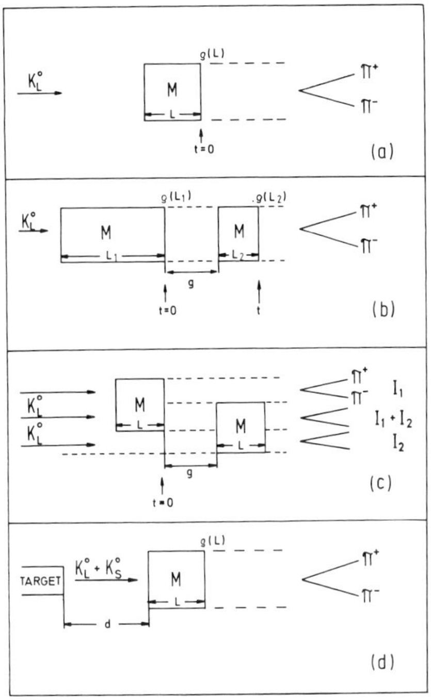


Fig.13 Methods for determining $\Delta m$. a) "regeneration" method, b) "gap" method, c) "zero-cross" method, d) $K_S^0 K_S^0$ interference method

effect was given in Sec.2.4). Four experimental setups using regenerated $K_S^0$ are described here separately.

Method R1 - Measurement of the $\pi^+\pi^-$ decay rate behind a piece of matter (regenerator) which is penetrated by a $K_L^0$ beam ("regeneration" method, see Fig.13a). This time-dependent rate is described as a function of the thickness L of the regenerator as given by (2.46) so that

$$R_{\pi^+\pi^-}(t,L) \propto \Big\{ |\rho(L)|^2 \cdot e^{-\lambda_S t} + |\eta_{+-}|^2 \cdot e^{-\lambda_L t} + 2|\rho||\eta_{+-}| \cdot \cos(\phi_\rho - \phi_{+-} + \Delta m t) \cdot e^{-(\lambda_S+\lambda_L)t/2} \Big\} \ .$$

The mass difference $\Delta m$ is then determined from the contribution of the interference term. This method was successfully applied in obtaining the first $|\Delta m|$ measurement [36]. But it is not so useful for getting precise measurements mainly because it relies too much on an accurate knowledge of the other parameters appearing in the expression of the decay rate, like regeneration parameters, CP violation parameters, etc.

Method R2 - Measurement of the $\pi^+\pi^-$ decay rate behind two pieces of regenerator separated by an air gap ("gap" method, see Fig.13b). The total rate of $\pi^+\pi^-$ decays results here from the addition of the decay amplitudes from the two pieces of regenerator. With $L_1$ and $L_2$ being the thicknesses of the pieces the rate behind the second piece is approximately given by

$$R_{\pi^+\pi^-}(t,g,L_1,L_2) = |\langle 2\pi|K(t)\rangle|^2 \propto \Big\{ |\rho(L_1)|^2 \cdot e^{-\lambda_S t} + |\rho(L_2)|^2$$

$$+ 2|\rho(L_1)||\rho(L_2)| \cdot e^{-\lambda_S t/2} \cdot \cos\left(\phi_{\rho(L_1)} + \phi_{\rho(L_2)} + \Delta m t\right)\Big\} . \qquad (4.1)$$

The time t is the time of flight (in the kaon rest system) for the distance $(g+L_2)$. For clarity the expression in (4.1) does not include the smaller contributions, as they come from the $(K_L^0 \to \pi^+\pi^-)$ amplitude, the regeneration in air, etc. (but they have of course to be included in the analysis of more precise experiments).

It follows from (4.1) that the measured rate is very sensitive to the gap length between the two pieces of regenerator and the mass difference. When g increases it decreases faster than the rate expected only as a result of decays over the distance g. One can say that there is a precession of the relative phases of the two $K_S^0$ amplitudes determined by $|\Delta m|$. Thus the mass difference can be obtained by comparing the rates for different separations g, if all the other parameters are adequately known.

It should be noted that the correlation with the regeneration parameters r (defined in (2.44a)) is here less important than for Method R1, since only the ratio $\rho(L_1)/\rho(L_2)$ is contributing (as far as the terms in (4.1) are concerned). The dependence on a knowledge of the CP violating amplitudes $(K_L^0 \to \pi^+\pi^-)$ is also weaker; its contribution to the decay rate is about 5% (see [20c]). The limitation in accuracy of $|\Delta m|$ measurement seems at present to be given by another correlation, namely that with $\lambda_S$, the reciprocal $K_S^0$ lifetime. Though it is possible to determine $\Delta m$ and $\lambda_S$ simultaneously via (4.1) one usually prefers to fix $\lambda_S$ to a value from independent measurements, for instance from the data tables [12].

Method R3 - Measurement of the $\pi^+\pi^-$ decay rate behind two pieces of regenerator arranged in such a way that the interference term of (4.1) is zero ("zero cross" method). This is a special case of the "gap" method. As shown in Fig.13c, one uses two pieces of matter of equal thickness ($L_1 = L_2 = L$, $\phi_{\rho(L_1)} = \phi_{\rho(L_2)}$) and compares the sum of the rates in track 1 and track 3 with the rate in track 2. The interference term of (4.1) obviously does not contribute when the compared numbers are equal. In this case one has $|\Delta m| \cdot t = \pi/2$ so that $|\Delta m|$ can directly be calculated from the measurement of the time t via distance $(g + L)$ and kaon momentum. This method does not depend on an accurate knowledge of the $K_S^0$ lifetime [55].

Method R4 - Measurement of the $\pi^+\pi^-$ decay rate behind a piece of regenerator with an incident beam of $K_S^0$ and $K_L^0$ (see Fig.13d). The main contribution to the $\pi^+\pi^-$ rate comes here from the superposition of two $K_S^0$ amplitudes, the incident one and the one produced by regeneration from the $K_L^0$ amplitude. The relative phase between the amplitudes is again proportional to $|\Delta m|$, but one has in addition an absolute reference for the phase expressed by

$$\cos\left(\phi_\rho - \frac{\Delta m d}{\lambda_S \Lambda_S}\right)$$

with $\Lambda_S$ being the scattering length (for details see [4]). Thus $\Delta m$ can be determined in magnitude and sign. With this method the identification of the sign turned out to be positive (i.e., $m_{K_L^0} > m_{K_S^0}$) [34]. But this method is unimportant for a precise measurement of $|\Delta m|$.

ABRAGAM [11] discussed an idea of driving the transition between $K_L^0$ and $K_S^0$ in matter by a modulation of the coherent regeneration. In the experiment the mass difference could be measured in an analogous way as one measures precisely the energy splitting of the two lowest states of the hydrogen atom due to the famous method of Lamb and Retherford. To make the regeneration of the $K_S^0$ amplitude resonant (and thus to perform the analogue) one must find a way to modulate at a suitable frequency the coherent regeneration amplitude (2.45b). It was shown by Abragam that an "unorthodox" use of a polarized target could provide the modulation via the spin dependence of the forward scattering amplitude f(o). However, this spin dependence is not known so that there is so far no quantitative basis for proposing experiments to determine the mass difference with this method.

Besides the methods measuring the mass difference $|\Delta m|$ by means of regenerated $K_S^0$, there is another class of methods of practical importance which is based on the "strangeness oscillation" in a neutral kaon beam already mentioned in Section 2.1. The oscillation is observed in the time dependence of $K^0$ and $\bar{K}^0$ intensity, respectively, as can be seen from (2.22b) and Fig.1. Three methods in this class are interesting on the experimental side.

Method 01 - Measurement of $\bar{K}^0$ intensity as a function of time in an initially pure $K^0$ beam by detecting strong interactions of $\bar{K}^0$ in matter. The $\bar{K}^0$ interactions are identified by the final states with the strangeness number $S = -1$, i.e., one looks for simple production of $K^-$, $\Lambda^0$, $\Sigma^0$ and $\Sigma^+$ or for the simultaneous production of $\Xi^0$ $K^+$, etc. The $\bar{K}^0$ intensity, due to (2.22b) mainly given by

$$R_{\bar{K}^0}(t) \propto \left\{1 + e^{-\lambda_S t} - 2\, e^{-(\lambda_S+\lambda_L)t/2} \cos\Delta m t\right\}$$

oscillates with time because of the cosine term. Periodicity as well as development in magnitude are dependent on $|\Delta m|$. The initial rise of this term is most sensitive to the magnitude of $|\Delta m|$ [6].

The analysis supposes a complete measurement of the double scattering process, the production process of the kaon followed by the secondary interaction. The best device for this complicated type of experiment might be the bubble chamber which simplifies problems by providing pictures, and where both scattering processes are on the same frame and available for a detailed analysis (see [37]). But on the other hand there is the disadvantage that one cannot easily obtain high event statistics.

In an interesting variant of this method one selects out of the secondary interactions of $K^0$'s and $\bar{K}^0$'s the final state $K^0_S p$. When measuring first, the kaon time of flight before the scattering process, t, and second, the decay time of flight of the $K^0_S$, t', one finds for the rate [35]

$$R(t,t') \propto \left\{\left|(A+B)\, e^{-\lambda_S t/2} + (A-B)\, e^{-(\lambda_L+2i\Delta m)t/2}\right|^2 \cdot e^{-\lambda_S t'}\right\} . \qquad (4.2)$$

Here A and B are the complex scattering amplitudes for the $K^0$ and $\bar{K}^0$ strong interactions in hydrogen. They have to be known from experiments with incident neutral kaon beams. The interesting point is that with this type of experiment $\Delta m$ could be measured in magnitude and sign simultaneously. It was in fact possible to confirm the positive sign in this way, but the accuracy in measuring $|\Delta m|$ is still very much limited by the lack of precise data from $K^0$ and $\bar{K}^0$ strong interactions at higher energies (see for example [35]).

Method 02 - Measurement of the time-dependent decay rate for semileptonic final states ($\pi^+\ell^-\nu$, $\pi^-\ell^+\bar{\nu}$). As can be seen in the expression (2.34), the $\Delta m$ dependence occurs again in the interference term producing the oscillation. This method is easier to realize than 01, but it has the disadvantage that one has to measure $|\Delta m|$ together with $x_\ell$, the parameter describing the violation of the $\Delta S = \Delta Q$ rule (see Sec.2.2), if one wants to obtain the same level of accuracy as in the experiments based on regenerated $K_S^0$. In a variant of this method, successfully applied recently [20b], the mass difference $|\Delta m|$ is determined from the charge asymmetry of $\pi\ell\nu$ decays expressed by (2.41). But also in this case $|\Delta m|$ and $x_\ell$ are correlated with each other (and with Re$\varepsilon$).

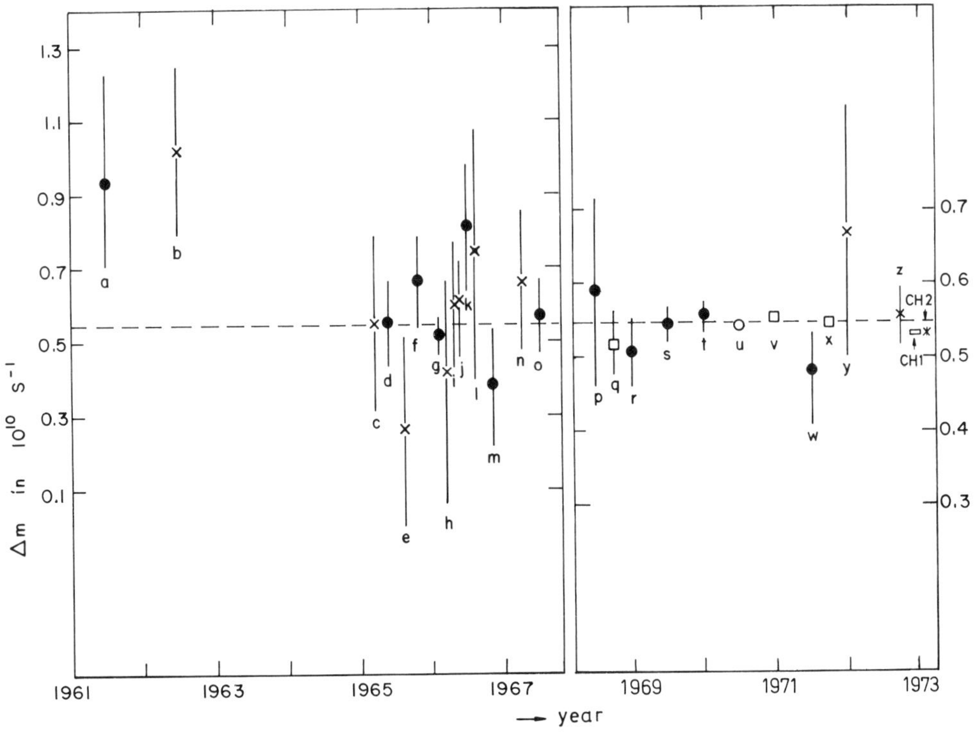


Fig.14 Measurements of $\Delta m$ in chronological order. The method is indicated as follows: •: R1; ×: 01 or 02; □: R2, o: R3. The references are: a to f [36-41], g [25], h to x [42-56], y [35], z [102], CH1 [20c], CH2 [20d]. The dashed line corresponds to the mean value of results exclusively CH1 and CH2

## 4.2 Experimental Results

The measurements of $|\Delta m|$ are shown in Fig.14 in a chronological order. The method applied is indicated in each case. As far as the data are dependent on the value of the $K_S^0$ lifetime $\tau_S$, they are presented here for the new value $\tau_S^{NEW} = 0.895 \cdot 10^{-10}$ s, i.e., the original results were approximately corrected with respect to the difference between the lifetime value assumed in the original calculation and this new value. From the experiments where $|\Delta m|$ and $\tau_S$ were determined simultaneously the data are taken without correcting them. The result from the "zero cross" method (called u in Fig.14) is, of course, unchanged, too. For the most precise experiments to be corrected (these are v and x) the shift in $|\Delta m|$ due to the effective increase in $\tau_S$ results in an increase of about 2% which corresponds, in terms of the errors (published with the original $|\Delta m|$ values), to two standard deviations.[8]

Table 2 Measurements of the mass difference $\Delta m = m_{K_L^0} - m_{K_S^0}$. The experimental methods are indicated by the labels introduced in Section 4.1

| Method | Experiment | Weighted mean value ($10^{10}\,\hbar\,s^{-1}$) | Calculated probability |
|---|---|---|---|
| R1 | [36], [39], [41], [25], [45], [47], [49], [50], [52], [53], [54], [57] | 0.543 ± 0.013 | 0.58 |
| R2 + R3 | [51], [55], [56], [58], [20b] | 0.541 ± 0.005 | 0.10 |
| O1 + O2 | [37], [38], [40], [42], [43], [44], [46], [48], [35], [102], [20c] | 0.534 ± 0.004 | 0.67 |
| R1 + R2 + R3 + O1 + O2 | | 0.5389 ± 0.0022 | 0.47 |

Table 2 contains the mean value of all the measurements collected in Fig.14 together with the mean values from subsamples which one gets by separating the

[8] Experiment v: $\Delta m^{ORIGINAL} = (0.542 \pm 0.006) \cdot 10^{10}\,\hbar\,s^{-1}$,
$\Delta m^{NEW} = 0.552 \cdot 10^{10}\,\hbar\,s^{-1}$

Experiment x: $\Delta m^{ORIGINAL} = (0.534 \pm 0.007) \cdot 10^{10}\,\hbar\,s^{-1}$,
$\Delta m^{NEW} = 0.547 \cdot 10^{10}\,\hbar\,s^{-1}$

data with respect to the methods applied. We distinguish three subsamples defined by the "regeneration" method R1, by the "gap" method R2 with "zero cross" method R3 and by the "strangeness oscillation" methods O1 and O2, respectively. The following conclusions can be drawn from Table 2 and Fig.14:

1) The corrected value of $|\Delta m|$ is mainly determined by the results of five experiments. These are (Fig.14) the points (crosses) indicated by u, v, x, CH1 and CH2. Two of them are obtained with method R2, two with R3 and one with O2.

2) These main results and the mean values for the corresponding subsamples are fairly consistent with each other. The overall mean value is obtained with a probability of 0.47.

3) The consistency of the data inside a subsample is not better than that between the subsamples.

4) Explanations of the apparent low level of confidence in the $|\Delta m|$ value are not directly obvious. New measurements might be needed in order to settle this problem, because the present status of $|\Delta m|$ measurements is not very satisfactory for the determination of $\phi_{+-}$, the phase of the CP violating $\pi^+\pi^-$ decay amplitude. This is discussed more in Section 5.2.

# 5. Measurement of CP Violation in the Two-Pion Decay Modes

## 5.1 Experimental Situation for Decays into $\pi^+\pi^-$ and $\pi^0\pi^0$

As explained in Section 2.2 the characteristics of CP violation to be determined for the two-pion decays are the magnitude and the phase of the complex parameters $\eta_{+-}$ and $\eta_{00}$. There are two basically different experimental methods that can be applied; either one measures the magnitude and the phase in a $K_S^0$ - $K_L^0$ interference experiment as explained in the context of (2.28), or one measures the magnitude directly in a $K^0$ beam by the decay rate ($K_L^0 \to 2\pi$) relative to a known $K_L^0$ decay, e.g., ($K_L^0 \to \pi^+\pi^-\pi^0$). These methods have also been used to measure relative values, i.e., the ratio of the magnitudes $|\eta_{+-}/\eta_{00}|$ and the difference of the phases ($\phi_{+-} - \phi_{00}$), which are both interesting to compare with theoretical predictions.

### 5.1.1 Results for $|\eta_{+-}|$ and $|\eta_{00}|$

The measurements of $|\eta_{+-}|$ are compiled in Fig.15. For each result the method used is indicated. One can see that the measurements done before 1972 are systematically lower than the more recent measurements; the older measurements,

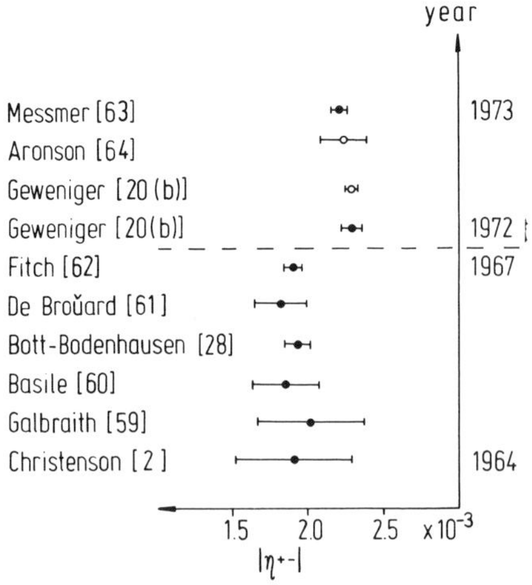


Fig.15 Measurements of $|\eta_{+-}|$ in chronological order. The experiments with direct measurement of the $(K_L^0 \to \pi^+\pi^-)$ rate are indicated by ●, the vacuum regeneration experiments by ○

which are in good agreement with each other, yield a value of $(1.98 \pm 0.04) \cdot 10^{-3}$ [12b]. The recent measurements, and in particular the results of the CH collaboration [20b] are about 15% higher.

During the Batavia conference in 1972, RUBBIA [64] discussed possible error sources in detail without finding an explanation for the large differences within the range of instrumental problems. For the results before 1972 which are all based on the measurement of the $R(K_L^0 \to \pi^+\pi^-)/R(K_L^0 \to \text{three-body state})$, several other partial decay rates of $K_L^0$ and $K_S^0$ are involved in the final calculation of $|\eta_{+-}|$. Rubbia pointed out that the difference in $|\eta_{+-}|$ of 15% means for the partial decay rates of $K_L^0$ and $K_S^0$ a difference of up to 30%. It is not very probable that the partial decay rates were measured with such a large systematic error. The only conclusion one can draw at the moment is that there might be some unrecognized systematic effects. One can say in favor of the new results (after 1967) that they come from both decay-rate measurements and interference experiments.

For the results from interference experiments the dependence of the result on the $K_S^0$ lifetime $\tau_S$ is of great importance: The change of $\tau_S$ to a higher value (see Sec.3) corresponds to a growth of $|\eta_{+-}|$ of about 0.2 [64], which is just in the order of magnitude of the difference between old and new $|\eta_{+-}|$ measurements.

The parameter $|\eta_{00}|$ was determined by the measurement of the decay-rate ratio

$$\frac{R(K_L^0 \to 2\pi)}{R(K_S^0 \to 3\pi)} \; .$$

The result averaged over all experiments is $|\eta_{00}| = (2.2 \pm 0.25) \cdot 10^{-3}$ [12c]. The accuracy of this result is not sufficient to allow a discrimination between the controversial $|\eta_{+-}|$ results. The ratio

$$\frac{|\eta_{00}|}{|\eta_{+-}|} = ( \ .0 \ 5 \pm 0.046) \qquad (5.1)$$

was measured in two recent experiments [65,66]. The result shows that $|\eta_{00}|$ and $|\eta_{+-}|$ are equal down to a few percent.

### 5.1.2 Results for $\phi_{+-}$ and $\phi_{00}$

The phase $\phi_{+-}$ is determined by the measurement of the $K_S^0 - K_L^0$ interference in one of two ways: either in the original $K^0$ or $\bar{K}^0$ beam described by (2.28), or by regenerated $K_S^0$ behind a block of material which was penetrated by a $K_L^0$ beam as described in the simplest case by (2.46).

For the determination of $\phi_{+-}$ the dependence on the other parameters also contained in the time distribution is very important. In the first case, the so-called vacuum regeneration, the number of parameters is smaller than for the second case because the regeneration parameters are not present. Nevertheless, the main problem of close interdependence with the mass difference $\Delta m$ is the same, because it is given by the form of the interference term $\sim |\eta_{+-}| \cos(\Delta m \cdot t - \phi_{+-})$. One has to know $\Delta m$ to determine $\phi_{+-}$ (or preferably to take into account the parallel measurement of the semileptonic decay distribution, which contains an interference term $\sim\cos(\Delta m t)$ (see Sec.2.2) as described for the CH experiment [20d]). The correlation between $\phi_{+-}$ and $\Delta m$ is very strong for all three vacuum regeneration experiments [20e,67,68]: $\phi_{+-}$ grows by $2^\circ$ to $3^\circ$ for a one percent increase of $|\Delta m|$. The details of the interdependence are illustrated in Table 3 (the values were computed via the $\Delta m$ dependence given in the references).

One sees that for the measuring the accuracy reached up to now a precise knowledge of $|\Delta m|$ is crucial. The results of the three experiments are compatible with each other for all values of $|\Delta m|$. A direct comparison has to be made with care, as the parameters $|\eta_{+-}|$ and $\tau_S$ were not fixed to the same values; these parameters also influence the determination, but certainly not as strongly as does $\Delta m$. Especially for the CH experiment the values measured in parallel were taken, which deviate rather strongly from the older measurements (see above). For the determination of $\phi_{+-}$ via the regeneration method

Table 3 Experimental results of $\phi_{+-}$

| Experiment | I $\Delta m = 0.547$ | II $\Delta m = 0.534$ | III $\Delta m = 0.538$ |
|---|---|---|---|
| [67] | $(51 \pm 12)^\circ$ | $(45 \pm 12)^\circ$ | $(47 \pm 12)^\circ$ |
| [68] | $(45 \pm 5)^\circ$ | $(40 \pm 5)^\circ$ | $(43 \pm 5)^\circ$ |
| [20e] | $(53.4 \pm 2.5)^\circ$ | $(45.9 \pm 1.6)^\circ$ | $(48.5 \pm 2.1)^\circ$ |

It is assumed for the mass difference $\Delta m$ that it is

I: the mean value without the CH results [20c,d]

II: the mean value of the two CH results only

III: the mean value of all measurements

The $\Delta m$ values are given in units of $10^{10}\,\hbar\,s^{-1}$.

[39,69,49,70,71,54,57,72], one has in principle the same problems with the dependence of $\phi_{+-}$ on $\Delta m$, $\tau_S$ and $|\eta_{+-}|$. The accuracy of these measurements has not yet reached the same level as the vacuum regeneration experiments, so that the problems discussed above are of less importance. The average value of the experiments is, with the average value of $|\Delta m|$ of all experiments (III)

$$\phi_{+-}^{RG} = (39.9 \pm 3.9)^\circ \ .$$

The result is compatible with that of the vacuum regeneration.

The average value of all $\phi_{+-}$ measurements is, with $\Delta m$ (III)

$$\phi_{+-}^{W} = (46 \pm 1.7)^\circ \ .$$

This result has to be interpreted with care as long as the value of $\Delta m$ is not known with more confidence.

As an alternative to the consideration of the mean values we will build up our further discussion in Section 9 on the completely consistent set of results of the CH experiment. The measurements $\phi_{00}$ are for the time being much less accurate than the $\phi_{+-}$ result because of greater experimental difficulties (detection of several $\pi^0$'s). The status of the $\phi_{00}$ measurements can be given by the mean value of the experiments [73,74]

$$\phi_{00} = (43 \pm 19)^\circ \ .$$

A first attempt to determine $\phi_{00}$ relative to $\phi_{+-}$ by a comparison of the decay distribution into $\pi^+\pi^-$ and $\pi^0\pi^0$ behind the same regenerator [75] yielded the result

$$(\phi_{00} - \phi_{+-}) = (7 \pm 18)^\circ \ . \qquad (5.2)$$

## 5.2 Isospin Analysis

Which part of the CP violating decays leads to the isospin 2 state of the two pions ($K^0$, $\bar{K}^0 \to 2\pi$, $I = 2$)? This important question can be answered in principle by means of the connection given in (2.27), if one uses the experimentally determined parameters $\eta_{+-}$ and $\eta_{00}$ and $\omega$. With the reasonable approximation $\omega \ll 1$ [6], one gets from (2.27)

$$\eta_{+-} = \varepsilon_0 + \varepsilon' \quad \text{and} \quad \eta_{00} = \varepsilon_0 - 2\varepsilon' \ . \qquad (5.3)$$

With the results for $|\eta_{00}/\eta_{+-}|$ and $(\phi_{00} - \phi_{+-})$ given in (5.1) and (5.2) one gets the following estimate for the ratio of the amplitudes $(\varepsilon'/\varepsilon)$ [64]:

$$\mathrm{Re}(\varepsilon'/\varepsilon) = 0.00 \pm 0.002 \quad \text{and} \quad \mathrm{Im}(\varepsilon'/\varepsilon_0) = 0.05 \pm 0.10 \ .$$

A more general analysis which has been performed recently [76,20e] was also based on (5.3). The following experimental data were used:

1) $|\eta_{+-}|$ from [20b]
2) $\phi_{+-}$ from [20e]
3) $|\eta_{00}/\eta_{+-}|$ as given in (5.1)
4) $\mathrm{Re}\,\varepsilon_0$ derived from the charge asymmetry of semileptonic decays [20a]
5) $\phi_{\varepsilon'}$ measured via the $\pi\pi$-phase shift analysis at the mass of the neutral kaon [77].

The relation $\phi_{\varepsilon'} = \pi/2 + \delta_2 - \delta_0$ holds with $\delta_2$ and $\delta_0$ being the "phase shifts" for the $I = 2$ and $I = 0$ wave, respectively. The results of the analysis were the following:

$$|\varepsilon_0| = (2.32 \pm 0.05) \cdot 10^{-3} \ , \quad |\varepsilon'/\varepsilon_0| = 0.007 \pm 0.015 \ , \quad \text{and}$$

$$\phi_{\varepsilon_0} = (45.4 \pm 1.3)^\circ \ .$$

This again shows that the contribution of the $I = 2$ amplitude to the CP violation in the two-pion decays is compatible with zero and cannot be greater than a few percent of the $I = 0$ amplitude. This result favors these theoretical models which contain $\varepsilon' = 0$.

# 6. Search for CP Violation in the Three-Pion Decay Modes

## 6.1 Experimental Situation for Decays into $\pi^+\pi^-\pi^0$

The experimental data for measuring CP violation in $\pi^+\pi^-\pi^0$ decays are compiled in Table 4. The reaction which produces the neutral kaons and the technique applied are indicated. Table 4 also gives the number of decays analyzed and the estimate of $\eta_{+-0}$, the parameter describing the contribution of $K^0_S \to \pi^+\pi^-\pi^0$ ($\eta_{+-0}$ was defined in Sec.2.2). The experiments add up to about 1200 $\pi^+\pi^-\pi^0$ decays where half of this number comes from bubble-chamber experiments, the other half comes from experiments with counter techniques. The experimental results, i.e., the $\eta_{+-0}$ values, are all consistent with zero; but it is certainly not straightforward to deduce for the single experiments upper limits of $\mathrm{Re}\eta_{+-0}$ and $\mathrm{Im}\eta_{+-0}$, because one can show that the error distributions (for statistical reasons) are still not gaussian-like. The problem was discussed in the study of $\pi^+\pi^-\pi^0$ decays by the COS collaboration [19b]. In this experiment $\eta_{+-0}$ was estimated from 180 $\pi^+\pi^-\pi^0$ decays which are very probably representing a uniquely identified and complete sample (see also Sec.3.1). Examining the behavior of the likelihood as a function of $\eta_{+-0}$ using (2.31) (i.e., applying the maximum likelihood method in the usual way, see for instance [79]) they obtained the result shown in Fig.16a. There are two maxima in the plane spanned by $\mathrm{Re}\eta_{+-0}$ and $\mathrm{Im}\eta_{+-0}$ which are surrounded by the likelihood contours relative to the corresponding maximum. The form of the contours indicates that the experiment cannot easily distinguish between a broad range of values of $\eta_{+-0}$ running from about (-2.0) to about zero through small negative values of $\mathrm{Im}\eta_{+-0}$.

The flow of the contours in Fig.16a was studied with simulations. As a first step the decay distributions given by (2.11) were calculated for specific values in the complex $\eta_{+-0}$ plane assuming constant detection efficiency. Fig.3 shows the time-dependent decay rates in comparison with each other. It can be seen that for $\eta_{+-0}$ values in an area which is defined in Fig.16a, roughly in the "banana" shape surrounded by contour 3, the distributions are

Table 4 Compilation of $\eta_{+-0}$ measurements

| N. | Exp. | Production process | Technique used | Observed number of $\pi^+\pi^-\pi^0$ decays | Result $\mathrm{Re}\,\eta_{+-0}$ | $\mathrm{Im}\,\eta_{+-0}$ |
|---|---|---|---|---|---|---|
| 1 | [80] | $\pi^- p \to K^0 \Lambda^0$ | HBC | 18 | $0.25 \pm 0.55$ | $-0.80 \pm 0.55$ |
| 2 | [81] | $K^+ n \to K^0 + X^+$ | HBLC | 190 | - | $0.34 {+0.19 \atop -0.59}$ |
| 3 | [82] | $K^- p \to \bar{K}^0 n$ | HBC | 53 | $0.05 \pm 0.30$* | $-0.15 \pm 0.45$* |
| 4 | [83] | $K^- p \to \bar{K}^0 n$ | HBC | 50 | $2.75 {+0.65 \atop -0.60}$ | $0.5 {+0.70 \atop -0.55}$ |
| 5 | [84] | $K^+ d \to K^0 pp$ | DBC | 99 | $0.47 {+0.30 \atop -0.24}$ | $-0.14 {+0.44 \atop -0.45}$ |
| 6 | [85] | $\pi^- p \to K^0 \Lambda^0$ | SC | 99 | $-0.09 \pm 0.19$ | $0.56 \pm 0.43$ |
| 7 | [86] | $K^+ p \to K^0 p \pi^+$ | WC | 384 | $0.13 {+0.17 \atop -0.20}$ | $0.17 {+0.27 \atop -0.26}$ |
| 8 | [19b] | $K^+ p \to K^0 p \pi^+$ | HBC | 180 | $0.17 {+0.17 \atop -0.23}$ | $0.01 {+0.38 \atop -0.40}$ |
| 9 | [87] | $\pi^- Cu \to K^0 + X^-$ | SC | 148 | $-0.05 \pm 0.17$ | $0.39 {+0.35 \atop -0.37}$ |
| 1-9 | | | | 1221 | | |

HBC = hydrogen bubble chamber; DBC = deuterium bubble chamber; HLBC = heavy-liquid bubble chamber; SC = spark chamber setup; WC = wire chamber setup.

* This result was obtained by addition of the data of [80] and [82].

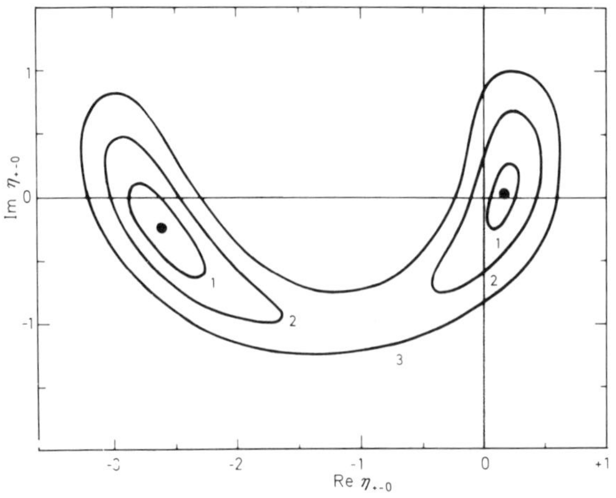


a) The result of the CERN-Oslo-SACLAY experiment. The numbers correspond to the standard deviations relative to the main maximum (at $Re\eta_{+-0} \simeq -2.7$)

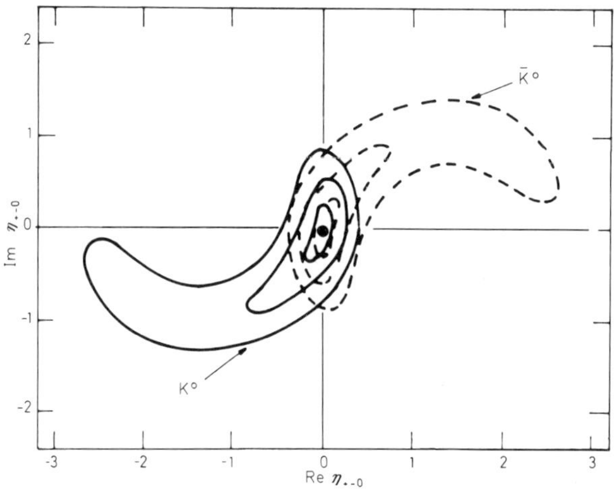


b) Simulated result for $\eta_{+-0} = 0$, assuming the experimental conditions of the CERN-Oslo-SACLAY experiment (full line). The dashed line shows the corresponding calculation assuming an originally pure $\bar{K}^0$ beam

Fig.16 Likelihood maxima and contours of equal likelihood for $\pi^+\pi^-\pi^0$ decays (from [19b])

mostly similar to that for $\eta_{+-0} = 0$ and may be easily simulated by statistical fluctuations. For the other values, however, the distributions have more structure which makes them more easily distinguishable from the $\eta_{+-0} = 0$ case. In particular for $Re\eta_{+-0} > 0$ there is a pronounced peak at short times due to the fact that the $\eta_{+-0}$ and $Re\eta_{+-0}\cos\Delta mt$ terms in (2.31) are added; the absence of such a peak in the experimental data allows such $\eta_{+-0}$ values to be ruled out with ease whereas the cases with $Re\eta_{+-0} < 0$, where the two terms can cancel, are less easily eliminated.

The possibilities of distinguishing the decay rate distributions for $\eta_{+-0} = 0$ from the others were also estimated quantitatively by simulating the experimental conditions of the COS experiment (for details see [19b,e]). Fig.16b shows the simulated contours corresponding to 1, 2 and 3 standard deviations relative to a maximum at $\eta_{+-0} = 0$ (full lines). The similarity to Fig.16a is evident. This contour shape has also been noticed in previous experiments [80,81,84,85,87], where subsidiary maxima have sometimes occurred and have been disposed of by various arguments. An interesting consequence of this effect concerns experiments where $\bar{K}^0$ are generated instead of $K^0$. The interference term in (2.31) has for $\bar{K}^0$ the opposite sign than for $K^0$ so that one expects to get the "banana" reflected at $\eta_{+-0} = 0$. The simulations confirm this relation as is obvious from the dotted lines in Fig.16b, and from the actual experiments with $\bar{K}^0$ [82,83].

It follows that a combination of $K^0$ and $\bar{K}^0$ data allows a more accurate determination of the amplitude (this was already pointed out by WEBBER et al. [82]). In a simultaneous analysis of $K^0$ and $\bar{K}^0$ data where the number of decays is roughly the same, the bias of the individual estimator is eliminated. This analysis is not possible with the experimental data available, because there are not enough $\bar{K}^0$ experiments. According to Table 3 only 103 $\pi^+\pi^-\pi^0$ decays in the sample of more than 1200 come from experiments with $\bar{K}^0$ production. The apparent imbalance between $K^0$ and $\bar{K}^0$ statistics is certainly a consequence of the fact that there are dominating $K^0$ production reactions which can easily be measured (in the absence of further neutral particles), whereas such reactions are almost totally missing for $\bar{K}^0$ production. Typical reactions at low energies are (2.1) and (2.2). The determination of the neutral kaon momentum (necessary for the identification of $\pi^+\pi^-\pi^0$ decays) is of course safest when one measures all secondaries. This is straightforward for (2.1), but requires for (2.2) the measurement of the neutron with high confidence. In the light of the experimental situation it would be desirable to have another generation of experiments measuring the time dependence of $\pi^+\pi^-\pi^0$ decays. Already with some $10^3$ $\pi^+\pi^-\pi^0$ decays, half from $K^0$ and half from $\bar{K}^0$ production, the ambiguities could be removed and the statistical error of $\eta_{+-0}$ would be decreased to the order of $10^{-2}$. For bubble-chamber techniques this means that one would have to analyze some $10^5$ $V^0$'s. In the case that one day one finds $|\eta_{+-0}| \neq 0$ on the $10^{-2}$ level, one has to answer the subsequent question of whether this is a proof of CP violation or only the measure of decays into higher angular momentum states (see Sec.2.2). A distinction is possible in the Dalitz plot of the decay pions, as has already been attempted [86].

In order to obtain a preliminary estimate of $\eta_{+-0}$, the data in Table 4 were used to calculate a mean value assuming gaussian error distributions [78,87]. The results are in agreement with

$$|\eta_{+-0}| < 0.3 \ . \qquad (6.1)$$

Assuming CPT invariance and the validity of the $\Delta I \leq 3/2$ rule of the isospin, GLASHOW and WEINBERG showed that $\mathrm{Re}\,\eta_{+-0}$ should be approximately zero when the amplitudes $(K_S^0 \to \pi^+\pi^-\pi^0)$ and $(K_L^0 \to \pi^+\pi^-\pi^0)$ are of comparable magnitude [89]. With $\mathrm{Re}\,\eta_{+-0} = 0$ one finds from the imaginary part

$$|\eta_{+-0}| < 0.2 \ . \qquad (6.2)$$

## 6.2 Experimental Situation for Decays into $\pi^0\pi^0\pi^0$

A first measurement of $\eta_{000}$, the parameter for estimating the contribution of CP violating $K_S^0$ decays into $\pi^0\pi^0\pi^0$, was carried through by a group in Moscow [88]. In this experiment the neutral kaons were produced with a separated $\pi^-$ beam of 3.5 GeV/c in a 180-liter ITEP xenon bubble chamber. The $\pi^0\pi^0\pi^0$ decays were detected on the pictures by the $e^+e^-$ pair production of the photons coming from $\pi^0$ decay. On $10^6$ pictures a total of 22 decays with six visible $e^+e^-$ pairs was identified. With these events which allow an unambiguous reconstruction of the decay process (6-constraint fit) the result is

$$\mathrm{Re}\,\eta_{000} = -0.04 \pm 0.45 \ , \quad \mathrm{Im}\,\eta_{000} = 0.45 \begin{array}{l} +\ 0.50 \\ -\ 0.65 \end{array} . \qquad (6.3)$$

We may note that this estimate is biased in the same sense as the $\eta_{+-0}$ results, namely that the assumption of gaussian errors does not hold. Neglecting this problem it follows from (6.3)

$$|\eta_{000}|^2 \leq 1.5 \ . \qquad (6.4)$$

Assuming $\mathrm{Re}\,\eta_{000} = 0$ [89] the result is

$$|\eta_{000}|^2 \leq 1.2 \; . \tag{6.5}$$

(All upper limits are quoted for "90% conf. level").

# 7. Test of the $\Delta S = \Delta Q$ Rule in the Semileptonic Decay Modes

## 7.1 Results for Decays into $\pi^- e^+ \bar{\nu}$ and $\pi^+ e^- \nu$

The most precise decay measurement was obtained by NIEBERGALL et al. [99]. They based their analysis on ~5,000 events of initial $K^0$ production according to (2.1) using a $K^+$ beam of 2.4 GeV/c momentum. A schematic view of their apparatus is shown in Fig.17. A hodoscope system is used to select $K^+$ interactions in the hydrogen target which produce four charged particles: two particles at large angles (proton and pion) in a cylindrical hodoscope around the target (TH) and two particles at small angles (the charged decay products of the neutral kaon) in hodoscope planes downstream to the target (DH I, DH II

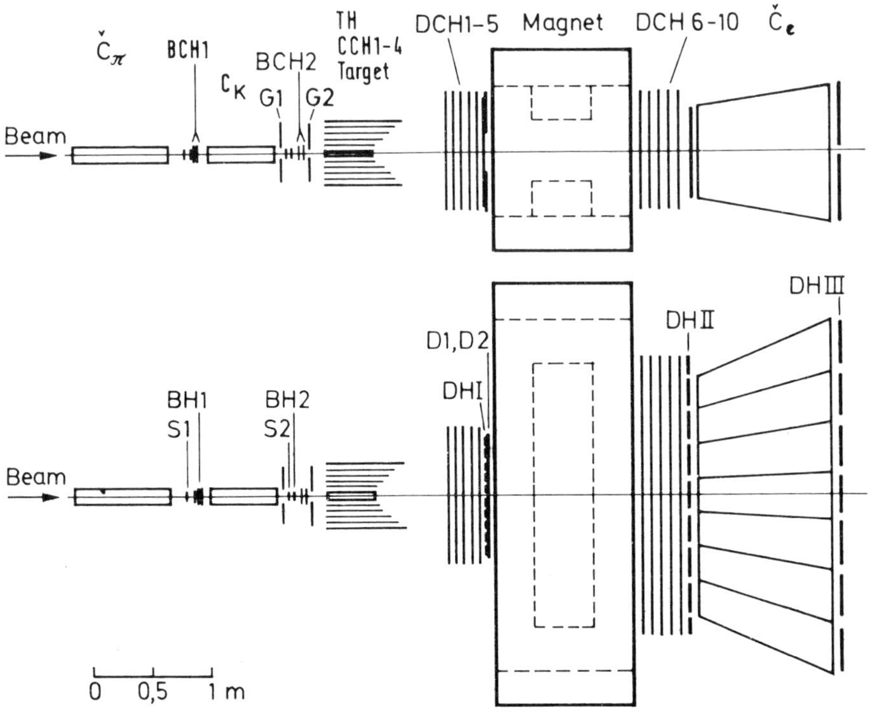


Fig.17 Plan view and elevation of the apparatus of NIEBERGALL et al. [99]. (Explanations are in the text)

and DH III). The directions of the four charged particles are measured with wire spark chambers (CCH and DCH). The decay particles are momentum analyzed in a wide-aperture magnet and electrons are identified with a Cerenkov counter (Ce). The time distributions of observed $\pi^- e^+ \bar{\nu}$ and $\pi^+ e^- \nu$ decays corrected for acceptance are shown in Fig.18. The full lines correspond to the predictions for $\chi_e = 0$. The best fit due to (2.34) and (2.35a) is obtained for

$$\mathrm{Re}\chi_e = 0.04 \pm 0.03 \;, \quad \mathrm{Im}\chi_e = -0.06 \pm 0.05 \;.$$

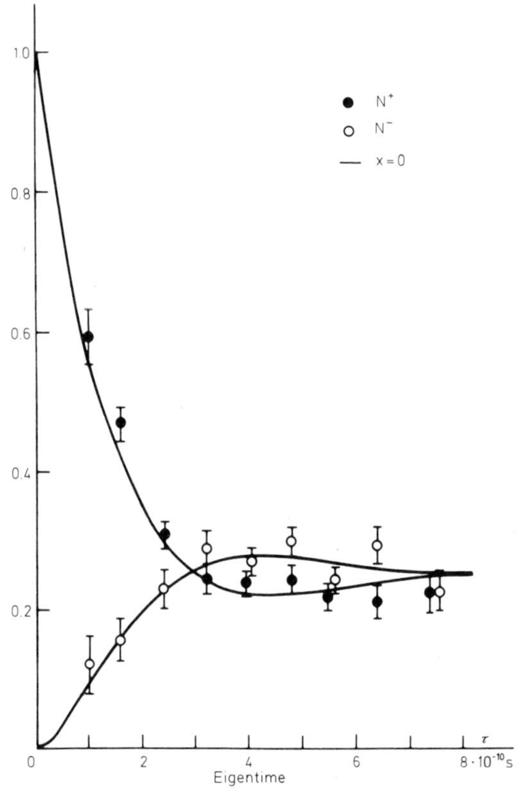


Fig.18 Proper time distribution of $\pi^- e^+ \bar{\nu}$ and $\pi^+ e^- \nu$ decays observed in the experiment of NIEBERGALL et al. [99]. The curves drawn in are the predictions for $\chi_e = 0$

Figure 19 gives a survey of all ($\Delta S = \Delta Q$) experiments, in which the complex parameter $\chi_e$ was determined via (2.34) of Section 2.3. The experimental equipment used is indicated in Fig.19; experiments in which the $\pi e \nu$-decays were not distinguished from $\pi \mu \nu$ in the same beam are labelled with an asterisk (these experiments have determined $\chi_\ell$ instead of $\chi_e$). The result of BENETT et al. [71] is the only one in this figure that was not derived via an interference measurement using (2.34), but via the measurement of charge

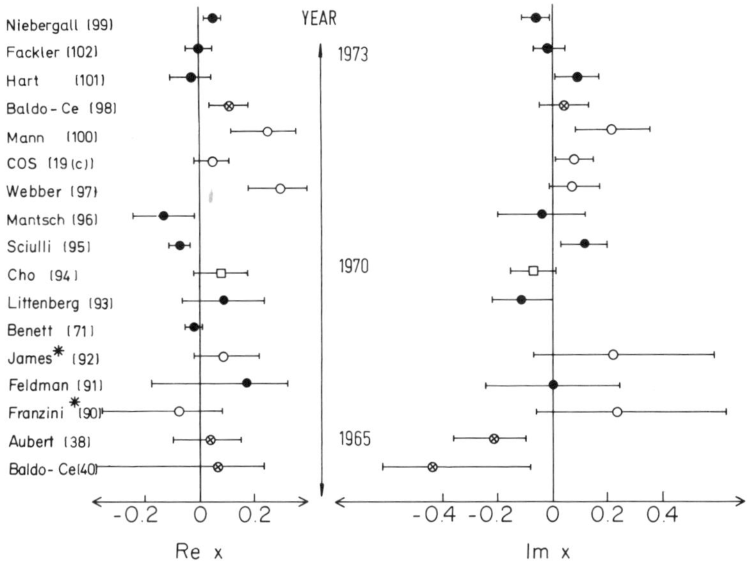


Fig.19 Measurements of $\chi_e$ for $\pi^-e^+\bar{\nu}$ ($\pi^+e^-\nu$) decays. Experiments which have not distinguished $\pi e\nu$ from $\pi\mu\nu$ are indicated by an asterisk. The experimental technique is indicated as follows: ● spark chambers; ○ hydrogen bubble chamber; □ deuterium bubble chamber; ⊗ heavy-liquid bubble chamber

asymmetry using (2.42), and yields therefore $\mathrm{Re}\chi_e$. One can see in Fig.19 that the more recent measurements are all very consistent with each other; the results are all compatible with zero and have an error of a few percent only. This is also true for the mean value

$$\mathrm{Re}\chi_e^W = 0.025 \pm 0.02 \ , \qquad \mathrm{Im}\chi_e^W = 0.008 \pm 0.016 \ .$$

This means that the $\pi e\nu$ decays at the present state are compatible with a strict validity of the $\Delta S = \Delta Q$ rule. At the same time because of $\mathrm{Im}\chi_e \simeq 0$, hypothetical CP violating contributions to $\pi e\nu$ decays are constrained inside a few percent.

In the experiments, measuring $\chi_e$ in the time distribution of the $\pi e\nu$ decay rate the data were always integrated over the range of the decay variables with rather arbitrary integration limits which depended on the possible range of electron identification. In bubble-chamber experiments electrons are identified by bubble density and therefore only for low-energy electrons, whereas in counter experiments, usually with Cerenkov counters, the decays with the fastest electrons (in the forward direction) dominate. Therefore the samples studied are in general not the same as far as the decay kinematics are concerned. With respect to the search for ($\Delta S = -\Delta Q$) contributions they are equivalent only if the ($\Delta S = -\Delta Q$) amplitude is proportional to the ($\Delta S = \Delta Q$) amplitude (or identical to zero). An investigation over the complete decay Dalitz plot is probably the next step. However, in order to proceed in this direction one might need a hybrid system of a bubble chamber and counters, where the small systematic errors and the $4\pi$ acceptance of the bubble chamber are combined with an efficient external electron identifier.

The $x_e$ values of Fig.19 are obtained assuming the validity of CPT symmetry; as explained in Section 2.3 this means the determination of the same parameter $x_e$ from both charge states $\pi^+e^-\nu$ and $\pi^-e^+\bar{\nu}$ in coupled analysis. The reason for this is, as already mentioned, usually the small number of events which does not allow a further subdivision. Nevertheless in one experiment [103] an additional analysis without the postulate of CPT symmetry could be made. The result is used to estimate the parameter of CPT violation (see Sec.8).

## 7.2 Results for Decays into $\pi^-\mu^+\bar{\nu}$ and $\pi^+\mu^-\nu$

The investigation of the $\Delta S = \Delta Q$ rule in the time distributions of the $\pi\mu\nu$ decays is much more difficult than the corresponding investigation of the $\pi e\nu$ decays because of the problem of efficient muon identification. A background of wrongly analyzed events as small as 1% could falsify the experimental results significantly [19c]. Up to now $\pi\mu\nu$ decays have been investigated in four experiments with less than 100 events each. The experimental results are shown in Fig.20. They are all compatible with $x_\mu = 0$. The mean values are

$$\mathrm{Re}\,x_\mu^W = 0.09 \begin{array}{l}+\,0.06\\-\,0.09\end{array}, \qquad \mathrm{Im}\,x_\mu^W = 0.05 \begin{array}{l}+\,0.1\\-\,0.09\end{array}.$$

They are also satisfactorily compatible with zero. The errors are still three times larger than those of $x_e^W$. Therefore conclusions about a difference between $x_e$ and $x_\mu$ which would establish $\Delta S = -\Delta Q$ by another form factor (expressed by y in (2.37) in Sec.2.3) still cannot be drawn. A possible improvement of $\pi\mu\nu$ experiments could be expected from a combination of a bubble chamber with an efficient external muon identifier. They could be planned in the near future because bubble chambers in the large laboratories (like BEBC in CERN) are being equipped with large-scale muon detectors.

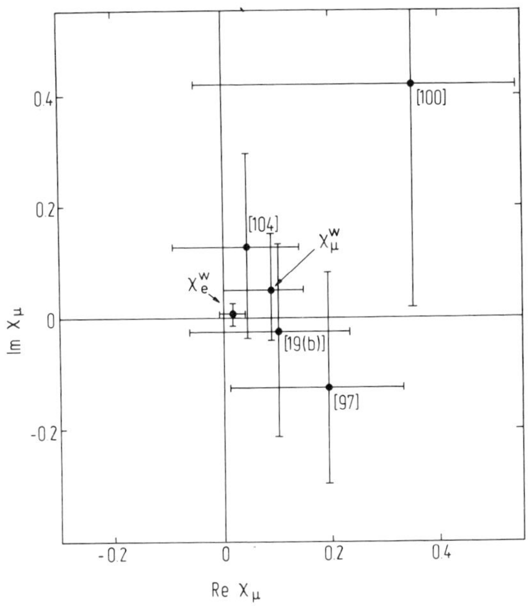


Fig.20 Measurements of $\chi_\mu$ for $\pi^-\mu^+\bar{\nu}$ $(\pi^+\mu^-\nu)$ decays. The mean values of $\pi\mu\nu$ and $\pi e\nu$ are the points indicated by $\chi_\mu^W$ and $\chi_e^W$, respectively

## 7.3 Charge Asymmetry Measurements

From $\mathrm{Re}\,\varepsilon \neq 0$ follows a charge asymmetry in the semileptonic decays, whose numeric value is changed if $\chi_e \neq 0$ (see (2.41) in Sec.2.3). The existence of the asymmetry has been proven in many experiments as well as measured with increasing accuracy. The results of the single experiments can be found in the compilation of the "Particle Data Group" [12c]. The mean values are:

$$\Delta_e = 0.326 \pm 0.023 \text{ for } (K_L^0 \to \pi e\nu) \text{ decays} \quad \text{and}$$

$$\Delta_\mu = 0.349 \pm 0.017 \text{ for } (K_L^0 \to \pi\mu\nu) \text{ decays} .$$

The mean values $\Delta_e$ and $\Delta_\mu$ are in good agreement with each other. If one assumes the validity of the $\Delta S = \Delta Q$ rule, which is supported by the interference experiments and by the asymmetry measurements (see, e.g., [20a]), one can calculate from a combination of $\Delta_e$ and $\Delta_\mu$ using (2.42) the real part of the CP violation parameter $\varepsilon$ with the result [12c]:

$$\mathrm{Re}\,\varepsilon = (1.72 \pm 0.10) \cdot 10^{-3} .$$

# 8. Analysis of the CP Violation Data Considering Unitarity

## 8.1 Formulation of the Unitarity Condition by Bell and Steinberger and Its Application

BELL and STEINBERGER have derived a sum rule for the conservation of the probability in the neutral kaon decays, also called unitarity condition, without assuming symmetry with respect to CP, T or CPT invariance [7]. Each neutral kaon state $|K(t)\rangle$ is a superposition of the eigenstates $K_S^0(t)$ and $K_L^0(t)$, so that one can write with (2.20) and (2.21):

$$|K(t)\rangle = a|K_S^0\rangle \cdot e^{-iE_S t} + b \cdot |K_L^0\rangle \cdot e^{-iE_L t}$$

with $E_{S,L} = m_{S,L} - \frac{i}{2}\lambda_{S,L}$ .

This state has a norm, whose time derivative at time $t=0$ is given by

$$-\frac{d}{dt}\langle K(o)|K(o)\rangle = \lambda_S|a|^2 + \lambda_L|b|^2 - 2\mathrm{Re}\{a^*b\langle K_S^0|K_L^0\rangle i(E_S^* - E_L)\} .$$

The shrinkage of the norm, which is given by this derivative, is caused by the decays; and because of conservation of probability it must be equal to the square of the sums of all decay amplitudes at time $t=0$

$$-\frac{d}{dt}\langle K(o)|K(o)\rangle = |a|^2 \cdot \sum_n |\langle K_n|K_S^0\rangle|^2 + |b|^2 \cdot \sum_n |\langle K_n|K_L^0\rangle|^2$$

$$+ 2 \cdot \mathrm{Re}\left\{a^*b\sum_n \langle K_n|K_S^0\rangle^* \langle K_n|K_L^0\rangle\right\} .$$

The comparison of the two equations can be made for any number a and b; therefore the coefficients must be equal

$$\lambda_{S,L} = \sum_n |\langle K_n|K_{S,L}^0\rangle|^2 \qquad (8.1)$$

and

$$i(E_L - E_S)\langle K_S^0 | K_L^0 \rangle = \sum_n \langle K_n | K_S^0 \rangle^* \langle K_n | K_L^0 \rangle \ . \tag{8.2}$$

Equation (8.2) is the Bell-Steinberger relation. It shows up an interesting correlation between the overlap of the two states $K_S^0$ and $K_L^0$ with the products of the transition amplitudes into common final states $K_n$. By comparison of (8.2) and (2.19) one can see that the non-orthogonality of $K_S^0$ and $K_L^0$ (i.e., $\langle K_S^0 | K_L^0 \rangle \neq 0$), which followed in Section 2.1 from the CP non-invariance of the H operator, is traced very generally to the existence of $K_S^0$ and $K_L^0$ decays into common final states. One can write with $\varepsilon$ and $\delta$ as defined in Section 2.1

$$2\,\mathrm{Re}\varepsilon + 2i\,\mathrm{Im}\delta = \frac{i}{E_S - E_L} \sum_n \langle K_n | K_S^0 \rangle^* \langle K_n | K_L^0 \rangle \ . \tag{8.3}$$

Equation (8.3) can be used in order to assess $\varepsilon$ and $\delta$ the parameters which analyze the CP violation in the Hamilton operator, in terms of T and CPT violation as described in Section 2.1. The sum on the right-hand side contains products of amplitudes which can be rewritten in terms of the parameters for measuring CP violation and violation of $\Delta S = \Delta Q$ rule in interference experiments. Measurements corresponding to two-pion, three-pion, and semileptonic decays have been discussed in the previous sections. In the sum the contributions from the two-pion decays are favored over the three-body decays by phase space. If one separates the dominating two-pion state with isospin zero (see Sec.5.2), one gets from (8.3)

$$\sum_n K_n | K_S^0 \rangle^* \langle K_n | K_L^0 \rangle = \varepsilon_0 \Gamma_0 + R \tag{8.4}$$

with the definitions

$$\varepsilon_0 = \frac{\langle 2\pi(0) | K_L^0 \rangle}{\langle 2\pi(0) | K_S^0 \rangle} \quad \text{and} \quad \Gamma_0 = |\langle 2\pi(0) | K_S^0 \rangle|^2 \ . \tag{8.5}$$

The letter R stands for all the other terms in the sum considered in (8.3). The contribution from the two-pion states with isospin zero is about $2 \cdot 10^{-3}$ $\Gamma_0$ (see isospin analysis, Sec.5.2).

In order to account for the leading role of the $(I=0)$ two-pion state, WU and YANG [6] have used the special choice of a real amplitude $A(0)$ for the $K^0$ transition into the $I=0$ state of the *standing* wave (for discussion of this condition see [6]). If CPT invariance holds, the corresponding $\bar{K}^0$ amplitude is also real; so under these conditions one finds for the transition amplitudes into the *outgoing* waves

$$<2\pi(0)|K_S^0> = \sqrt{2}\, A(0)\, e^{i\delta_0} \,, \qquad <2\pi(0)|K_L^0> = \varepsilon \cdot \sqrt{2} \cdot A(0)\, e^{i\delta_0} \,.$$

This means that we have in the Wu-Yang gauge

$$\varepsilon_0 = \varepsilon .$$

In order to estimate R, the contributions from decays, into states other than the $I=0$ two-pion state, we take here the invariance under CPT for granted. (In this way only data analyzed with this assumption can be included). We refer to an analysis without this assumption in Section 8.2.

The contributions to R, which have to be taken into account then, can be expressed by

$$R = i[2 \cdot \mathrm{Re}A(2) \cdot \mathrm{Im}A(2) - \Gamma_{\pi^+\pi^-\pi^0} \mathrm{Im}\eta_{+-0} - \Gamma_{3\pi^0} \mathrm{Im}\eta_{000} - 2 \cdot \Gamma_{\pi\mu\nu} \cdot \mathrm{Im}\chi_\mu -$$

$$- 2 \cdot \Gamma_{\pi e\nu} \cdot \mathrm{Im}\chi_e] \,. \qquad (8.6)$$

The parameter $A(2)$ denotes the decay amplitude into the $I=2$ two-pion state; the parameters $\eta_{+-0}$, $\eta_{000}$ are defined in Section 2.2, $\chi_\mu$ and $\chi_e$ in Section 2.3. (An explicit derivation of (8.6) is given for example by FAISSNER [18]). The contribution of the $I=2$ two-pion state is given by

$$<2\pi(2)|K_S^0><2\pi(2)K_L^0> = 2 \cdot \Gamma_0 \cdot \omega \cdot \varepsilon'$$

(see Sec.2.2 for definitions of $\omega$ and $\varepsilon'$). Since $\omega \ll 1$ and $\varepsilon' \ll \varepsilon_0$ it is of much less importance than the corresponding contribution of the $I=0$ state.

One can again get the expression for the $I=2$ state contained in (8.6) by analyzing this contribution in terms of the transition amplitudes of $K^0$ and $\bar{K}^0$, following WU and YANG [6].

The three-body decay modes contribute to (8.6) through the imaginary parts of corresponding parameters which are multiplied by the partial decay widths and are therefore suppressed by more than three orders of magnitude compared to the $I=0$ two-pion state. If one uses the experimental results described in Sections 5, 6 and 7, one obtains with (8.6) for the sum of the small contributions approximately

$$R = i\,0.7 \cdot 10^{-4}\,\lambda_S$$

with an error of the same order; therefore because $\lambda_S \simeq \Gamma_0$ in the Bell-Steinberger relation, R cannot be greater than a few percent of the main contribution by the $I=0$ two-pion state.

## 8.2 T Non-Invariance and CPT Invariance

The sum rule (8.2) can now be used to study the correlation of CP violation with T non-invariance and CPT non-invariance, respectively. Eq. (2.19) shows that $\langle K_S^0 | K_L^0 \rangle$ for CPT invariance (i.e., $\delta = 0$) is real and for T invariance is purely imaginary. Therefore an analysis of T and CPT invariance can be performed directly by the determination of the real and imaginary parts of the overlap element $\langle K_S^0 | K_L^0 \rangle$.

For CPT invariance one gets the following condition from (8.3):

$$\mathrm{Im}\left(\frac{i}{E_S - E_L}\right) \sum_n \langle K_n | K_S^0\rangle^* \langle K_n | K_L^0 \rangle = 0 \ .$$

With a little algebra one gets from this a direct correlation between the measured quantities for CP violation and the $K^0$ decays [105]

$$\sin(\phi_{\Delta m} - \phi_{+-}) + R_S \cdot |\eta_{00}/\eta_{+-}| \cdot \sin(\phi_{\Delta m} - \phi_{00})$$

$$= \frac{1 + R_S}{\lambda_S |\eta_{+-}|} \cdot \mathrm{Im} \sum_n \langle K_n | K_S^0\rangle^* \langle K_n | K_L^0 \rangle \cdot e^{i\phi_{\Delta m}}$$

with

$$\phi_{\Delta m} = \tan^{-1}\frac{\Delta m\tau_S}{h} \; ; \qquad R_S = \frac{\Gamma(K_S^0 \to \pi^0\pi^0)}{\Gamma(K_S^0 \to \pi^+\pi^-)} \; . \tag{8.7}$$

It is clear that the estimation of the left half of (8.7) is dominated by the error of $\phi_{00}$ (see Sec.5). With the latest measurements the result is LHS $= 0.0 \pm 0.2$. The right-hand side is also compatible with zero; the estimate based on the experimental results described in Sections 5, 6 and 7 is RHS $= 0.5 \pm 0.5$. The numbers are compatible with CPT invariance, but the estimate gives only a rather crude upper limit mainly because of the large measurement error on $\phi_{00}$.

The result agrees very well with estimates based on a general procedure due to SCHUBERT et al. to separate T and CPT non-invariant effects [103,106]. In their procedure the Bell-Steinberger relation (8.3) is used together with a second relation between $\varepsilon$ and $\delta$ which is obtained from the definition of $\varepsilon_0$ (8.5) by inserting (2.15) for small $\varepsilon$ and $\delta$:

$$\varepsilon_0 = \varepsilon - \delta + \alpha_0 \tag{8.8}$$

with

$$\alpha_0 = \frac{A(o) - \bar{A}(o)}{A(o) + \bar{A}(o)} \; ; \qquad \begin{aligned} A(o) \cdot e^{i\delta_0} &= \langle 2\pi | K^0 \rangle \\ \bar{A}(o) \cdot e^{i\delta_0} &= \langle 2\pi | \bar{K}^0 \rangle \end{aligned} \; .$$

The parameter $\delta_0$ denotes the $I = 0$ $\pi\pi$ phase shift. With a phase convention which gives $A(o)$ and $\bar{A}(o)$ the same phase (e.g., the Wu-Yang convention mentioned above), one gets $\mathrm{Im}\,\alpha_0 = 0$; for CPT invariance one gets $\alpha_0 = 0$.

The parameter $\varepsilon_0$ can be split up into two parts following (8.8); first, into $\varepsilon$ (T non-invariant and CPT invariant) and second, into

$$\tilde{\delta} = \delta - \alpha_0 \; .$$

The parameter $\tilde{\delta}$ is a T invariant and CPT non-invariant quantity. The two complex parameters $\varepsilon$ and $\tilde{\delta}$ can be determined from the experimental results by the two complex equations (8.3) and (8.8). Of course one can now use only the results determined without taking CPT invariance for granted (i.e., most of the measurements of the semileptonic decays compiled in Figs.19 and 20 cannot be used!). The result of the computation given by RUBBIA not too long ago [64] gives a quantitative characterization of the experimental situation

$$\mathrm{Re}\,\tilde{\delta} = (0.01 \pm 0.25) \times 10^{-3}\,, \qquad \mathrm{Im}\,\tilde{\delta} = (-0.24 \pm 0.30) \times 10^{-3}\,,$$

$$\mathrm{Re}\,\varepsilon = (1.39 \pm 0.25) \times 10^{-3}\,, \qquad \mathrm{Im}\,\varepsilon = (1.16 \pm 0.26) \times 10^{-3}\,. \tag{8.9}$$

This result is compatible with CPT invariance and T non-invariance. Upper limits for a deviation from CPT invariance can also be derived by another method based on the measurement of the decay rate integrated up to 6 $\tau_S$. A recent result [107] gives an upper limit of $10^{-3}$ for $\mathrm{Re}\,\delta$.

If CPT invariance holds, T invariance has to be violated along with the observed CP violation, i.e., the equation

$$\mathrm{Re}\left(\frac{i}{M_S - M_L}\sum_n \langle K_n|K_S^0\rangle^* \langle K_n|K_L^0\rangle\right) = 0$$

is not fulfilled.

STEINBERGER [105] rewrites this as follows:

$$\cos(\phi_{\Delta m} - \phi_{+-}) + R_S \cdot |\eta_{00}/\eta_{+-}| \cdot \cos(\phi_{\Delta m} - \phi_{00})$$

$$= -\frac{1 + R_S}{\lambda_S |\eta_{+-}|}\,\mathrm{Re}\sum_n \langle K_n|K_S^0\rangle^* \langle K_n|K_L^0\rangle\, e^{i\phi_{\Delta m}}\,. \tag{8.10}$$

The parameters $\phi_{\Delta m}$ and $R_S$ are defined as in (8.7); LHS = 1.4 ± 0.2. But the right-hand side gives RHS = 0.2 ± 0.2 which agrees with zero. The significance of this deviation shows up more clearly in the results of Schubert et al. where $\varepsilon$ is found to be nonzero by five standard deviations (see (8.9)).

The T non-invariance can be established directly by the measurement of the muon polarization in $K_L^0 \to \pi^\pm \mu^\mp \nu$. In the $K_L^0$ system a component of muon polarization normal to the decay plane is odd under time reversal T and forbidden by T invariance [5]. In a recent experiment [108] the muon polarization was measured by the method of spin precession with improved accuracy, but results are still non conclusive.

# 9. Possibilities of Explaining CP Violation

## 9.1 Classification of Sources

The total Hamilton operator, which includes all interactions (i.e., strong electromagnetic, weak and eventually an additional interaction) can be split into a CP invariance and a CP violating part so that

$$H = H_+ + H_- \quad \text{with} \quad CP\, H_\pm (CP)^{-1} = \pm H_\pm$$

The (CP = -1) interactions are described by $H_-$. From $K_L^0 \to 2\pi$, we know that it exists, but there are no positive indications from processes other than neutral kaon decays.

For obvious reasons one cannot immediately deduce from the existence of CP violating decays the interaction which causes the effect. Apart from the possibility that $H_-$ is an operator of the weak interaction with $\Delta Y = 1$[9], which produces the CP violating decays directly, there is also the possibility that $H_-$ is connected to the electromagnetic or strong interaction and because of $\Delta Y = 0$ leads indirectly to the decays through virtual CP violating processes.[10] Furthermore $H_-$ could be the effect of an additional interaction, which would only be detectable by the $K^0$ decays. The most important example for that is the so-called superweak interaction with $\Delta Y = 2$.

In order to select the most probable explanation(s), one has to look at the quantitative side; that is, one has to compare the predictions from models based on a definite source with the experimental results inside and outside the neutral kaon decays. The data on CP violation in ($K^0$, $\bar{K}^0$) decays are in fact rather inefficient for providing a decision; for all the hypothetical sources mentioned above there are some models in quantitative agreement with the data. Actually the experiments with other particles play a progressively more important role because for these one can prove or disprove quantitative predictions as a consequence of the CP violation in $K^0$ decays. These predictions in general differ by orders of magnitude because of the differences in strength with respect to the interactions responsible. So the models could be tested there with greater sensitivity. However as a matter of fact most

[9] The hypercharge Y is defined by baryon number B and strangeness S in the following way: $Y = B + S$. For the kaon because of $B = 0$ the hypercharge is equal to the strangeness quantum number, so $\Delta S$ and $\Delta Y$ are the same.

[10] For an illustration of such cases see [18].

of the possible experiments have not yet been performed with sufficient precision to enable a unique choice of the models to be made.

Important exceptions are the experiments which measure the electric dipole moment of the neutron. They have been improved very much recently and will probably be pushed to even higher accuracy in the near future [109,110]. An electric dipole moment defined by

$$\vec{d}_n = \langle n|e\vec{r}|n\rangle$$

can exist only if the neutron state is non-invariant under the parity transformation P as well as time reversal T (because it must vanish for each neutron state invariant under P or T). A neutron state with $d_n \neq 0$ would then have P- and T-violating admixtures

$$|n\rangle \quad |n_0\rangle + a|n_1\rangle$$

with a the coupling constant which describes the simultaneous P and T violation. The magnitude of the moment can then be estimated by

$$d_n = a \cdot e \cdot r$$

with r being the characteristic dipole length. Setting this length at the typical value of $10^{-14}$ cm, one obtains

$$d_n = a \cdot 10^{-14}\ e \cdot cm\ . \tag{9.1}$$

The present upper limit from the experiments is [110]

$$d_n^{exp} < 10^{-23}\ e \cdot cm\ .$$

From the comparison with (9.1) it follows then that

$$a < 10^{-9}\ .$$

Theoretical estimates of the constant a, as we shall see, can differ greatly depending on the model used to compute it. In the following we shall discuss the essential points of the different types of models, which give no clear contradiction to the measurements of the $K^0$ system. In Section 9.2 the case of H being a $(\Delta Y = 2)$ operator will be discussed. In Section 9.3 the agreement between the experimental results and the manifold models with $H_-$ as ($\Delta Y = 1$) operator or ($\Delta Y = 0$) operator will be investigated.

## 9.2 Description of Models on Superweak Interactions ($\Delta Y = 2$)

The superweak models first introduced by WOLFENSTEIN and LEE and WOLFENSTEIN [9], respectively, start from the following assumption:

$$H_+ = H_{ST} + H_{EL} + H_W \quad \text{and} \quad H_- = H_{SW} . \tag{9.2}$$

The Hamiltonians $H_+$ and $H_-$ are assumed to be CPT invariant. The term $H_W$ denotes the normal weak interaction, which carries $\Delta Y = 0$ or $\Delta Y = 1$ where Y is the hypercharge. The term $H_{SW}$ denotes the superweak interaction characterized by $\Delta Y = 2$. With the assumptions (9.2) the superweak interaction cannot directly lead to any observed $K^0$ decay, because there the hypercharge is only changed by one. However indirectly the superweak interaction changes the states of the real particles $K_S^0$ and $K_L^0$ by transitions like $\langle K^0|H_{SW}|\bar{K}^0\rangle$.

For the non-diagonal elements of the total energy matrix one obtains with the assumptions (9.2) and the definition (2.11) of Section 2.1:

$$H_{12} = \langle K^0|H_{SW}|\bar{K}^0\rangle + \langle K^0|(H_S + H_{SW})\,\frac{1 - P_0}{E - H_0}\,(H_S + H_{SW})|\bar{K}^0\rangle + \cdots$$

and

$$H_{21} = \langle \bar{K}^0|H_{SW}|K^0\rangle + \langle \bar{K}^0|(H_S + H_{SW})\,\frac{1 - P_0}{E - H_0}\,(H_S + H_{SW})|K^0\rangle + \cdots$$

$$= -\langle K^0|H_{SW}|\bar{K}^0\rangle + \cdots . \tag{9.3}$$

The connection with $\varepsilon$, the parameter of CP violation with simultaneous T non-invariance (and CPT invariance), can be expressed approximately using (2.16) and (2.14) in the following way [134]:

$$\varepsilon \simeq \frac{1}{2}\,\frac{H_{12} - H_{21}}{H_{12} + H_{21}} .$$

This expression looks as follows with (9.3) inserted:

$$\varepsilon \simeq \frac{1}{2} \frac{\langle K^0 | H_{SW} | \bar{K}^0 \rangle}{\langle K^0 | (H_S + H_{SW}) \frac{1 - P_0}{E - H_0} (H_S + H_{SW}) | \bar{K}^0 \rangle} . \tag{9.4}$$

Equation (9.4) means that $\varepsilon$ is non zero because of the superweak interactions, so that the $K^0$ eigenstates are correspondingly described by (2.15) for the case $\delta = 0$. Put the other way, one single parameter $\varepsilon$ already describes completely the CP violation in the $K^0$ decays caused by the superweak interaction. From (9.4) one can also get an idea of the strength of $H_{SW}$. The numerator of (9.3) is proportional to the coupling constant $G_{SW}$ of the superweak interaction; the denominator is proportional to the square of the coupling constant G of the normal weak interaction and the even weaker couplings $G \cdot G_{SW}$ or $G_{SW}^2$. So, for the order of magnitude of $G_{SW}$, one gets from (9.4)

$$G_{SW} = |\varepsilon| \cdot G^2 . \tag{9.5}$$

With $G = 10^{-5}/m_p^2$ (as usual in units of the reciprocal mass squared) one gets from (9.5), if one inserts the measured CP violation for the two-pion channels (with the approximation $|\varepsilon| = |\eta_{+-}|$):

$$G_{SW} \simeq 10^{-13} \, m_p^{-4} .$$

Expressed in terms of dimensionless coupling constants defined by

$$g = \frac{G \cdot m_p^2}{4\pi} \quad \text{and} \quad g_{SW} = \frac{G_{SW} \cdot m_p^2}{4\pi}$$

one gets (see also [4]):

$$g_{SW} = 10^{-9} , \quad g = 10^{-15} . \tag{9.6}$$

An estimate of $g_{SW}$ starting from the measured mass difference between $K_L^0$ and $K_S^0$ made by OKUN and PONTECORVO [111] yields the less restrictive result

$$g_{SW} \leq 10^{-7} \cdot g = 10^{-13}$$

which is nevertheless not in contradiction to (9.6).

From (9.6) one can see that the superweak interaction is about a thousand times weaker than the second-order weak interaction. Therefore processes with possible ($\Delta Y = \pm 2$) transitions such as the decays of the $\Xi$ into nucleon and pion are more likely to proceed via $H_W^2$ than via $H_{SW}$. The experimental upper limit for the corresponding $\Xi^0$ decay is given by SOERGEL [112] as

$$\frac{R(\Xi^0 \to p + \pi^+)}{R(\Xi^0 \to \Lambda + \pi^0)} < 3.5 \cdot 10^{-5}$$

which is still a few orders of magnitude larger than the rate to be expected from $H_W^2$.

The most interesting predictions of the superweak models are compiled in Table 5 and compared with experimental results.

Prediction 1 follows from the fact that the CP violation caused by the ($\Delta S = \pm 2$) transitions between $K^0$ and $\bar{K}^0$ has nothing to do with the transition matrix $H_W$. From the assumptions made and from (2.18) one gets immediately

$$<2\pi|H|K_L^0> = \varepsilon \cdot <2\pi|H_S|K_1^0> \quad \text{and} \quad <3\pi|H|K_S^0> = \varepsilon \cdot <3\pi|H_S|K_2^0> \ .$$

From that, one gets with the definitions introduced in Section 2.2:

$$\eta_{+-} = \eta_{+-0} = \eta_{000} = \varepsilon \ . \tag{9.7}$$

As can be seen in Table 4 the agreement for the magnitude of $\eta_{+-}$ and $\eta_{00}$ is confirmed experimentally down to the percent level. Moreover $|\eta_{+-0}|$ and $|\eta_{000}|$ cannot be checked for the time being with sufficient accuracy (for the detailed reasons see Sec.6).

The phase angles $\phi_{+-}$, $\phi_{00}$ have to be equal as well. They are expected to be equal to $\phi_{\varepsilon_0}$, too (to the degree that $\varepsilon'$ can be neglected). The angles $\phi_{+-}$ and $\phi_{\varepsilon_0}$ are determined with greater precision than $\phi_{00}$ (see Sec.5).

The Bell-Steinberger relation (8.2) is simplified by the assumptions (9.2) of the superweak theories as described in the following:

1) CPT invariance yields connected with (2.19):

$$<K_S^0|K_L^0> = 2 \cdot \mathrm{Re}\ \varepsilon \ . \tag{9.8}$$

Table 5  Predictions from superweak models

| | Prediction | Experimental result |
|---|---|---|
| 1 | $\lvert\eta_{+-}\rvert = \lvert\eta_{00}\rvert = \lvert\eta_{+-0}\rvert = \lvert\eta_{000}\rvert = \lvert\varepsilon\rvert$ | $\left\lvert\frac{\eta_{00}}{\eta_{+-}}\right\rvert = 1.02 \pm 0.05$ |
| 2 | $\Gamma(K_S^0 \to \pi^+\ell^-\nu) = \Gamma(K_L^0 \to \pi^+\ell^-\nu)$ | |
| | $\Gamma(K_S^0 \to \pi^-\ell^+\bar{\nu}) = \Gamma(K_L^0 \to \pi^-\ell^+\bar{\nu})$ | |
| 3 | $\frac{\Gamma(K_S^0 \to \pi^-\ell^+\bar{\nu})}{\Gamma(K_S^0 \to \pi^+\ell^-\nu)} = \left\lvert\frac{1+\varepsilon}{1-\varepsilon}\right\rvert^2$ | |
| 4 | $\phi_{+-}^{Th} = \phi_{00} = \tan^{-1}\frac{2\Delta m \cdot \tau_S}{\hbar}$ | $\phi_{+-}^{exp} = (46.0 \pm 1.7)^\circ$ * |
| | $= (44.0 \pm 0.2)^\circ$ * | $\phi_{00}^{exp} = (43 \pm 19)^\circ$ |
| 5 | $Re_\varepsilon^{Th} = \lvert\eta_{+-}\rvert \cdot \left\{1 + \left(\frac{2\Delta m\tau_S}{\hbar}\right)^2\right\}^{-1/2}$ | $Re\ \varepsilon^{exp} + (1.72 \pm 0.10) \cdot 10^{-3}$ |
| | $= (1.68 \pm 0.03) \cdot 10^{-3}$ * | |

For the values indicated by * it was taken into account that $\tau_S = 0.895 \cdot 10^{-10}$ s and $\Delta m = 0.5389 \cdot 10^{10}\ \hbar\ s^{-1}$. The $\phi_{+-}^{exp}$ value contains from the CH experiment $\phi_{+-}^{CH} = (48.5 \pm 1.8)^\circ$, which differs from the published value [20d] because of the $\Delta m$ value taken into account. The calculation of $Re_\varepsilon^{Th}$ is based on $\lvert\eta_{+-}\rvert = (2.20 \pm 0.04) \cdot 10^{-3}$ of the CH experiment [20b].

2) If there is no CP violation via $H_W$ the right-hand side of (8.2) is dominated by the I = 0 two-pion decays; i.e., for this part one gets, neglecting contributions of the other decays and using the approximation $\Gamma_0 = \lambda_S$

$$\sum_n \langle K_n|H|K_S^0\rangle^* \langle K_n|H|K_L^0\rangle = \varepsilon \cdot \lambda_s \,. \tag{9.9}$$

With the approximation

$$i(M_L - M_S) \simeq i\Delta m + 1/2\, \lambda_S$$

one gets from (8.2) together with (9.8) and (9.9) the relation

$$\left(2 \cdot i \cdot \frac{\Delta m}{\lambda_S} + 1\right) \lambda_S \cdot \mathrm{Re}\,\varepsilon = \lambda_S \cdot \varepsilon \ .$$

From this follow immediately the predictions 4 and 5 of Table 5. The computations for $\phi_{+-}$ and Re$\varepsilon$ made in this way are in agreement with the experimental results. The reliability of the comparison for $\phi_{+-}$ has to be discussed from the aspect that prediction as well as measurement depends on a knowledge of the mass difference $\Delta m$ between $K_L^0$ and $K_S^0$ and of the lifetime $\tau_S$ of the $K_S^0$. The error on the predicted value $\phi_{+-}^{Th}$ results immediately from the measuring errors of $\Delta m$ and $\tau_S$. As described in Section 5 the influence of $\Delta m$ and $\tau_S$ result also in the additional uncertainties on $\phi_{+-}^{exp}$ of the order of one to three degrees for a single experiment. Therefore the error for $\phi_{+-}^{exp}$ given in Table 5 has to be taken as a lower limit. Predictions 2 and 3 are put forward under the additional assumption of the validity of the $(\Delta S = \Delta Q)$ rule [9]. Up to now there are no appropriate data to check these predictions.

The comparison between predictions and experimental results of the $K^0$ decays has so far not led to any inconsistencies. One has to take into account, however, that the agreement is based mainly on the relatively more accurate results from the two-pion decays. Apart from the determination of Re$\varepsilon$ via charge asymmetry of the semileptonic decays the three-particle decays contribute only by results which exclude CP violations two to three orders of magnitude stronger than those of the two-pion decays. This is certainly not sufficient to prove the validity of a superweak model apart from the fact that the predictions for the two-pion channels are identical to the predictions of all those models which assume the ratio of isospin 2 to isospin 0 amplitude $\langle 2\pi(2)|K_S^0\rangle / \langle 2\pi(0)|K_L^0\rangle$ to be real. This condition implies that a CP violating amplitude is proportional to $\varepsilon$ (this can be seen from similar considerations to those described for the isospin zero amplitudes in Sec.8.1; see also [41]).

One also gets predictions in agreement with the experimental two-pion decays if one assumes CP violation in the two-pion channels to be the consequence of a maximal CP violation in the $\pi\ell\nu$ amplitude, as discussed by SACHS in connection with a corresponding violation of the $(\Delta S = \Delta Q)$ rule [113]. As far as the predicted maximal violation of the ($\Delta S = \Delta Q$) rule is concerned (see Sec.6), the Sachs model is in contradiction with the experiments.

The possibilities of detecting superweak interactions outside the kaon system are very restricted because of the extremely small strength of $10^{-9}$ compared with the usual weak interaction. The best chances to provide some insight are expected to be in connection with the measurement of an electric dipole moment of the neutron. The prediction of superweak models for the magnitude of the dipole moment yield extremely small values. Assuming a coupling constant $a = 10^{-15}$ in accordance with (9.6) one gets from (9.1)

$$d_n^{SW} = 10^{-29}\ e \cdot cm\ .$$

In the case that superweak interaction is restricted to processes which change strangeness ($\Delta S = 2$ as suggested by the neutral kaon decays) the electric dipole moment of the neutron cannot be produced directly. The corresponding estimates of $d_n^{SW}$ are than around $10^{-38}$ e · cm or less.

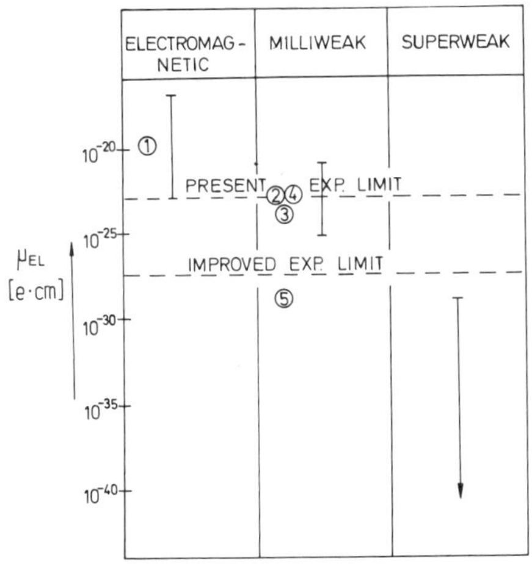


Fig.21 Predictions for the electric dipole moment of the neutron and experimental limits. The reference numbers 1 to 5 correspond to [115-119]

The predictions are compiled in Fig.21 following recent compilations [114]. In Fig.21 the range of superweak predictions is indicated together with milliweak and electromagnetic (with milliweak electromagnetic) predictions. Some recent predictions are included separately. The dashed lines mark the experimental results, the upper line the upper limit obtained so far [110], and the lower line the most advanced improvement supposed to be feasible (see BYRNE et al. [109]). One can conclude from Fig.21 that a direct test of superweak predictions is very unlikely in the near future. Indirectly the superweak predictions would be disproved if an electric dipole moment is detected at a smaller order of magnitude than $10^{-29}$ e · cm. On the other hand, superweak predictions will be strongly favored in the opposite case, that electromagnetic, milliweak or even millistrong predictions are disproved (i.e., $d_n^{exp} < 10^{-27}$ e · cm).

## 9.3 Description by Other Models

### 9.3.1 ($\Delta Y = 1$) Models

Many models for the description of CP violation start from the assumption that $H_-$ is a ($\Delta Y = 1$) operator. In that case $H_-$ has the same hypercharge quantum number as the normal weak interaction $H_W$. Therefore in this type of model weak processes in general have CP violating amplitudes. If one takes $K_L^0 \to \pi^+\pi^-$ as a typical example of CP non-invariance one gets a coupling constant F characteristically of $H_-$ about three orders of magnitude smaller than G, the coupling constant of $H_W$: $F/G = 10^{-3}$ based on $|\eta_{+-}| \simeq 10^{-3}$. Because of this ratio the interaction described by $H_-$ is called milliweak. The predictions of milliweak models [121] for CP violation outside the kaon system are much larger than those of the superweak models. Nevertheless experimental results of sufficient accuracy are not yet at hand to test these predictions. This will be explained in the following.

First of all, the experimental proof that the CP violation observed in the two-pion decays of the neutral kaon is also present with the same strength in the three-particle decays (see Secs.6 and 7) is missing. In hyperon decays the direct estimate of the degree of CP invariance is based on the comparison of the transition amplitudes for particles and antiparticles [4,5]. Experimental results of that kind do not yet exist. An appropriate experiment could be the simultaneous production of $\Lambda$ and $\bar{\Lambda}$ in $\bar{p}p$ interactions if one had a spectrometer with sufficiently large acceptance for $\Lambda$ and $\bar{\Lambda}$ detection.

In addition to particle/antiparticle comparison, there is the possibility of estimating the effects of CP violation in hyperon decays indirectly via the effects of T non-invariance, which could be seen by a comparison of the scattering phases of the amplitudes which determine the weak decay on one hand and the corresponding pion-baryon scattering on the other hand [4]. The necessary scattering phases are determined for $\Lambda$ decays with a relatively good accuracy [122,123]. Nevertheless the experimental errors are still too large to allow one to distinguish CP violating milliweak components by comparison with pion-baryon scattering phases (not much better known, see [124]). Even for nuclear β decays measured with much higher accuracy (as in an experiment with polarized $Ne^{19}$ beam [125]), one cannot yet get precise enough results to decide for or against milliweak T and CP violation (for a more detailed discussion see [4]). The most promising comparison between experiment and milliweak prediction is possible with respect to the electric dipole moment of the neutron, whose connection with CP violation was already mentioned in the previous sections. One can obtain a rough estimate from (9.1) with the milliweak coupling taken as

$$a = 10^{-3} \cdot G$$

so that

$$d_n^W = 10^{-23}\ e \cdot cm\ .$$

In Fig.21 one can see that the range of the "classical" milliweak predictions (discussed in detail by WOLFENSTEIN [121]) overlaps the experimental range so that the decision for or against these models is in progress. The predictions [116-119] are based on renormalizable and gauge invariant models. With one exception they also give $10^{-23}$ to $10^{-24}$ $e \cdot cm$. In the prediction of FRENKEL and EBEL [119] the dipole moment is suppressed by an additional factor $10^{-3}$. One can always get such a factor if one assumes the validity of a selection rule such as $|\Delta S| = 1$ for the CP violating part of the Hamilton operator $H_-$ (see [120]).

In this case $H_-$ cannot contribute directly to the electric dipole moment of the neutron, only terms of higher order in the coupling constant can contribute, i.e., terms of second order if a ($|\Delta S| = 1$) selection rule holds. It is clear, then, that a further confinement of the electric dipole moment of the neutron by more accurate experiments would in the first instance only contradict those milliweak theories which do not assume additional selection rules resulting in corresponding suppression factors.

### 9.3.2 ($\Delta Y = 0$) Models

These models start from the assumption that the CP violating part of the Hamiltonian $H_-$ is an operator acting on hadrons (but not on leptons) and fulfills the selection rule $\Delta Y = 0$. Furthermore, $H_-$ is assumed to be invariant under CPT and P, but not under C and T, and therefore not under CP. With these assumptions the CP violation in the ($K_L^0 \to 2\pi$) decays is described as an effect of second order which results from a CP invariant weak interaction $H_W$ and the C violating interaction $H_-$. The corresponding transition amplitudes are built up from terms of the form $H_W \cdot H_-$. Starting from such an interdependence, one can describe the kaon effects of CP violation such as $|\eta_{+-}| = 2 \cdot 10^{-3}$ correctly, if $H_-$ is much stronger than the usual weak interaction $H_W$ with a relative strength of $10^{-3}$. Electromagnetic as well as strong processes could lead to the correct strength for $H_-$. The existence of $H_-$ can in this case be detected by C and T non-invariance in electromagnetic or strong interactions. If one assumes an electromagnetic source with C violation for hadrons (which does not affect the C invariance of leptons) one expects in $K_L^0 \to 2\pi$ a CP violation in the order of $\alpha/2\pi \simeq 2 \cdot 10^{-3}$ multiplied by the CP invariant amplitude $K_S^0 \to 2\pi$, which would be a natural explanation for the effects observed in the decays of the neutral kaon.

The predictions for the kaon system coincide with those of the superweak models. Nevertheless an electromagnetic theory probably cannot be used for the description of CP violation, because the C non-invariance outside the kaon system predicted by this theory has not been seen experimentally. First of all, this is true for the electric dipole moment of the neutron which according to Fig.21 is predicted in general to be larger than the upper limit would allow. In a second case, charge asymmetry in the (G parity violating) electromagnetic decays of the $\eta^0$ into $\pi^+\pi^-\pi^0$ and $\pi^+\pi^-\gamma$, predicted on the same lines, is missing experimentally. The charge asymmetry in the $\eta^0$ decays is defined by

$$A = \frac{N(E\pi^+ > E\pi^-) - N(E\pi^- > E\pi^+)}{N(E\pi^+ > E\pi^-) + N(E\pi^- > E\pi^+)}$$

with N the number of events and $E\pi^\pm$ the pion energy. For the decays observed here there must be absolute symmetry between $\pi^+$ and $\pi^-$ if C invariance holds, i.e., A must be zero. On the other hand C non-invariance automatically leads to $A \neq 0$.

Theoretical estimates based on distinction between the possible isospin configurations of the three particles and on assumptions about centrifugal barriers yield values above $10^{-3}$ for $|A|$ in the case of $(\eta^0 \to \pi^+\pi^-\pi^0)$ decay, and in some cases larger values still (details in [4,126]). The average value of the relevant experiments is at the lower limit of the predictions and well compatible with zero, namely $A_{exp} = (1.2 \pm 1.7)\ 10^{-3}$ [12c]. For the $(\eta^0 \to \pi^+\pi^-\gamma)$ decay the prediction is $|A| \leq 1.5 \cdot 10^{-2}$ [4] compared to an experimental average [12c] of $A_{exp} = (8.8 \pm 4.0)\ 10^{-3}$. In both decays the experimental results are quite compatible with C invariance in electromagnetic processes.

Investigations of T non-invariance which would test CP violation more directly than the investigations of C invariance have led to negative results in agreement with the results concerning C invariance. In two experiments [127,128] in which T invariance was studied in the low-energy nuclear reactions,

$$^{24}Mg + a \rightleftarrows {}^{27}Al + p \quad [127] \qquad \text{and} \qquad ^{24}Mg + d \rightleftarrows {}^{25}Mg + p \quad [128]\ ,$$

by comparing the two possible directions of the reactions the maximal T non-invariant part was estimated to be less than $5 \cdot 10^{-3}$ relative to the whole reaction.

The alternative case in the $(\Delta Y = 0)$ models, namely $H_-$ to be given by C and T non-invariance in strong interactions, was also studied [9,130] and has been discussed in detail by CABBIBO [129]. The arguments for and against strong sources of the CP violation are much less clear than for the electromagnetic models. First of all it is not clear how to get the correct strength for the CP violation observed in the kaon decays in a natural way. The experimental results for the kaon decays point towards the so-called millistrong models which can give the same predictions for the kaons as the milliweak models. An example of the few developments in other directions is the model of PRENTKI and VELTMAN [130], which starts from a medium strong SU(3) breaking interaction.

On the experimental side it is mainly the C invariance in strong interactions which has been tested. A rather convenient possibility is the test of the charge asymmetries in the meson spectrum produced by $\bar{p}p$ annihilation. As was shown by PAIS [131], there are symmetries between equal particles of opposite charge if C or CP invariance holds for unpolarized $\bar{p}$ beam and unpolarized target protons independent of the explicit angular momentum state of the $\bar{p}p$ system. To test these symmetries one compares energy distributions, etc., of positive and negative kaons (pions) produced in the reactions

$$\bar{p} + p \to m\pi^+(K^+) + m\pi^-(K^-) + n\pi^0 \qquad m = 1,2,\cdots \qquad n = 0,1,\cdots .$$

The most significant results are those of BALTAY et al. [132] for antiprotons at rest and of DOBRZYNSKI et al. [133] for antiprotons with incident momentum of 1.26 GeV/c. Both experiments give a result $\leq 10^{-2}$ for the ratio between C non-invariant and C invariant amplitudes. It is not a straightforward matter to define a connection between this number and the CP violation observed in the kaon decays (see [4,126]), but a strong C non-invariance in strong interactions is already excluded.

## 10. Summary

The dominant decay processes of neutral kaons belong to two classes of weak interaction, the purely hadronic and the semileptonic class. From the experimental point of view, all the hadronic and semileptonic decays are of particular importance, since they allow $K_L^0 - K_S^0$ interference effects to be measured, even if they are caused by very small admixtures of certain symmetry breaking decay interactions.

In interference experiments of this kind, the violation of the CP symmetry is mainly investigated which was originally detected in the weak decay of the neutral kaons into a pair of charged pions. In the meantime CP violation has been investigated experimentally for all important decay modes of the neutral kaon, namely, in the decays into two pions, $\pi^+\pi^-$ and $\pi^0\pi^0$, and into three pions, $\pi^+\pi^-\pi^0$ and $\pi^0\pi^0\pi^0$, as well as in the semileptonic decays into the end states $\pi^\pm e\,\nu$ and $\pi^\pm\mu\,\nu$. For the semileptonic decays, the test of CP violation is usually made within the framework of the check of an important selection rule of weak interaction, the so-called $\Delta S = \Delta Q$ rule. This rule allows only those decays in which the transition from the neutral kaon to the charged pion changes strangeness and charge in the same way ($\Delta S = \Delta Q$).

The status of the interference experiments concerned with the four pionic and the four semileptonic decays of the neutral kaon is the following:

1) CP violation has been observed up to now only in the two-pion decays. The results are compatible with equal CP violation for the decays into $\pi^+\pi^-$ and $\pi^0\pi^0$. The relative magnitude of the CP violating amplitudes ($|\eta_{+-}|$ and $|\eta_{00}|$) is $2 \cdot 10^{-3}$, and the relative phases $\phi_{+-}$ and $\phi_{00}$ are both about $45^\circ$.

2) The measurements concerned with CP violation in the three-particle decays ($\pi^+\pi^-\pi^0$, $\pi^0\pi^0\pi^0$, $\pi^\pm e^\mp\nu$, $\pi^\pm\mu^\mp\nu$) are at the moment still one to two orders of magnitude less precise than the corresponding measurements of the two-pion decays (not counting the special case of the charge asymmetry measurements for semileptonic modes). With these measurements it is not possible to look for interference effects with amplitudes of a relative magnitude of $10^{-3}$ as observed in the two-pion channels.

3) The two-pion decays have recently been studied in improved experiments, especially with respect to the measurement of all parameters of the $\pi^+\pi^-$ decay, which are connected directly or indirectly with CP violation. Among these parameters are, on the one hand, the magnitude and phase of the $\eta_{+-}$ parameters, and on the other hand, the lifetime $\tau_S$ of the short-lived $K_S^0$ and the mass difference $\Delta m$ between $K_L^0$ and $K_S^0$. Essential parts of these new measurements show large discrepancies compared to the older experiments, so that neither the lifetime $\tau_S$ nor the mass difference $\Delta m$ nor the magnitude of the CP violating amplitude $|\eta_{+-}|$ is compatible with the corresponding mean values of the older measurements. Quite a few arguments can be found in favor of the new results, as detailed in Sections 3, 4 and 5. Nevertheless, for a final solution to these problems further measurements are certainly necessary. This would be especially important with respect to the phase $\phi_{+-}$ of $\eta_{+-}$, whose determination depends critically on other parameters, especially on the precise knowledge of the mass difference $\Delta m$.

4) The validity of the $\Delta S = \Delta Q$ rule is at present not questioned by the measurements of the semileptonic decays, but should be checked with greater accuracy by future experiments. In particular the measurements of the $\pi^\pm\mu^\mp\nu$ decays are not good enough yet to allow investigation of the relative importance of the different form factors by comparison with the $\pi^\pm e^\mp\nu$ decays.

The experimental results with respect to CP violation have been the bases for several interesting conclusions, which are certainly unchanged when considering the previous set of data instead of the present set. From the measured numbers that parametrize the CP violation, i.e., the relative amplitude $|\eta| \simeq 2 \dagger 10^{-3}$ and $\phi \simeq 45^\circ$ of the two-pion decays and the upper limits of $3 \cdot 10^{-1}$ and $10^\circ$ of the three-pion decays and $10^{-2}$ of $\pi^\pm\ell^\mp\nu$ decays ($\ell$ means e or $\mu$) it was concluded that CP violation has to be accompanied by some non-variance for time reversal (T), while the validity of CPT symmetry is not questioned (see Sec.8). Up to now, T non-invariance has not been detected elsewhere.

The experimental results from the neutral kaon decays cannot lead to an unambiguous explanation of CP violation, but they constrain the range of possibilities rather strongly, especially through the relatively precise results from the two-pion decays. The agreement between $|\eta_{+-}|$ and $|\eta_{00}|$ down to the $10^{-2}$ level and the $\phi_{+-}$ value of about $45^{\circ}$ together with the absence of an experimental proof of CP violation outside the kaon complex prompts one to believe in the superweak models, which connect CP violation to a new interaction of relative strength $10^{-9} \cdot G$ (G = coupling constant of weak interaction). Superweak effects have been observed nowhere else. In the light of the smallness of the strength, they are not detectable in other processes. One can also get similar descriptions of the decays of the neutral kaon with models based on known interactions, i.e., weak or electromagnetic or strong interactions. These models, in contrast to the superweak models, would also have consequences for processes other than the kaon decays, which can be tested against measurements in special experiments. Most important are the experiments which measure the electric dipole moment of the neutron. By decreasing the upper limit for this dipole moment, they have already led to the experimental exclusion of electromagnetic and most of the (milli-) weak models as explanation for CP violation. Further improved experiments are in preparation which might decide whether really superweak models finally remain.

## Acknowledgements

This article was written following my stay at CERN where I worked with Dr. L. Montanet on experimental problems in neutral kaon decays. I wish to thank him for this fruitful time, and I greatly appreciate the cooperation of all my colleagues in his group.

I am grateful to Prof. W. Paul for his support and continuing interest in this article, and I am indebted to Prof. H. Rollnik for critically reading parts of the manuscript and for making available to me the notes from his Seminar on $K^0$ Physics.

Thanks are due to Dr. H. Kowalski for many stimulating discussions. Also I thank Dr. K. Müller for assistance in preparing the final version and Dr. M. Mansfield for checking the English. Finally I would like to thank Mrs. A. Wasserziehr for the careful typing of the manuscript.

## References


1 L. Leprince-Ringuet, M. Lheritier: Comp. Rend. 219, 618 (1944).
2 J.H. Christenson, et al.: Phys. Rev. Letters 13, 138 (1964).
3 R.P. Feynmann, M. Gell-Mann: Phys. Rev. 109, 193 (1958).
4 R.E. Marshak, Riazuddin, C.P. Ryan: Theory of Weak Interactions in Particle Physics, Interscience Monographs and Texts in Physics and Astronomy 24 (J. Wiley, New York 1969).
5 L.M. Chounet, J.M. Gaillard, M.K. Gaillard: Phys. Reports 4C, 199 (1972).
6 T.T. Wu, C.N. Yang: Phys. Rev. Letters 13, 180 (1964).
7 J.S. Bell, J. Steinberger: Proc. Intern. Conf. Elementary Particles, Oxford (1965), p. 195.
8 T.D. Lee, C.S. Wu: Ann. Rev. Nucl. Science 16, 511 (1966); 17, 513 (1967).
9 T.D. Lee, L. Wolfenstein: Phys. Rev. 138B, 1490 (1965);
L. Wolfenstein: Nuovo Cimento 42A, 17 (1966).
10 L.B. Okun: Soviet Phys. Usp. 9, 574 (1967).
11 A. Abragam: Rapport DPh No. 1 (1968), CEN, Saclay.
12 Review of particle properties, Particle Data Group:
a) Phys. Letters 39B. 1 (1972);
b) Rev. Mod. Phys. 45, (1973);
c) Phys. Letters 50B, 1 (1974).
13 D. Banner, et al.: Phys. Rev. D7, 1989 (1973).
14 S. Eliezer, P. Slinger: Phys. Rev. 165, 843 (1968).
15 L.M. Sehgal: Phys. Rev. Letters 21, 412 (1968).
16 B.R. Webber: Lawrence Radiation Lab., Thesis (1969), unpublished.
17 A. Pais, O. Piccioni: Phys. Rev. 100, 1487 (1955).
18 P.K. Kabir: The CP Puzzle (Academic Press, London 1968). Detailed descriptions are given by H. Faissner, PITHA (1968), Rept. 26. A summarizing discussion of regeneration experiments before 1971 was given by K. Kleinknecht, invited talk prepared for the Daresbury Study Weekend, January 30-31, 1971.
19 CERN-Oslo-SACLAY Experiment:
a) O. Skjeggestad, et al.: Nucl. Phys. B48, 343 (1972);
b) F. James, et al.: Phys. Letters 35B, 265 (1971); Nucl. Phys. B49, 1 (1972);
c) G. Burgun, et al.: Nuovo Cimento Letters 2, 1169 (1971); Nucl. Phys. B50, 194 (1972);
d) G. Burgun, et al.: Phys. Letters 46B, 481 (1973).
e) G. Burgun, CEA-N-1702, Thesis (1974).
20 CERN-Heidelberg Experiment:
a) C. Geweniger, et al.: Phys. Letters 48B, 483 (1974);
b) C. Geweniger, et al.: Phys. Letters 48B, 487 (1974);
c) C. Geweniger, et al.: Paper 1082, 17th Intern. Conf. High Energy Physics, London (1974);
d) S. Gjesdal, et al.: Paper 1083, 17th Intern. Conf. High Energy Physics, London (1974);
e) C. Geweniger, et al.: Paper 1084, 17th Intern. Conf. High Energy Physics, London (1974).
21 J. Prentki: in Proc. 1962 Intern. Conf. High Energy Physics at CERN (1962), p. 839.
22 L. Bertanza, et al.: Phys. Rev. Letters 8, 332 (1962) and preprint (1962).
23 M. Chretien, et al.: Phys. Rev. 131, 2208 (1963).
24 M.N. Kreisler, et al.: Phys. Rev. 136B, 1074 (1964).
25 C. Alff-Steinberger, et al.: Phys. Letters 21, 595 (1966).
26 L. Auerbach, et al.: Phys. Rev. 149, 1052 (1966).

27 C. Baltay, et al.: Phys. Rev. 142, 932 (1966).
28 M. Bott-Bodenhausen, et al.: Phys. Letters 23, 277 (1966).
29 L. Kirsch, P. Schmidt: Phys. Rev. 147, 939 (1966).
30 R. Donald, et al.: Phys. Letters 27B, 58 (1968).
31 D.G. Hill, et al.: Phys. Rev. 171, 1418 (1968).
32 G. Ekspong, L. Voyvodic, J. Zoll: CERN Yellow Rept. 73-14.
33 D. Lord, E. Quercigh: DD./DH/70/20 and D.Ph. II/INSTR. 70/7.
34 W.A. Mehlhop, et al.: in Proc. Intern. Conf. High Energy Physics, Berkeley (1966).
35 D.G. Hill, et al.: Phys. Rev. D4, 7 (1971).
36 R.H. Good, et al.: Phys. Rev. 1223 (1961).
37 U. Camerini, et al.: Phys. Rev. 128, 362 (1962).
38 B. Aubert, et al.: Phys. Letters 17, 59 (1965).
39 V.L. Fitch, et al.: Phys. Rev. Letters 15, 73 (1965).
40 M. Baldo-Ceolin, et al.: Nuovo Cimento 38, 684 (1965).
41 I.H. Christenson, et al.: Phys. Rev. 140B, 74 (1965).
42 M. Baldo-Ceolin, et al.: Nuovo Cimento 45A, 733 (1966).
43 U. Camerini, et al.: Phys. Rev. 150, 1148 (1966).
44 C.Y. Chang, et al.: Phys. Letters 23, 702 (1966).
45 T. Fujii, et al.: Phys. Rev. Letters 13, 253 (1966).
46 G.W. Meissner, et al.: Phys. Rev. Letters 16, 278 (1966).
47 J.V. Jovanovich, et al.: Phys. Rev. Letters 17, 1075 (1966).
48 J.M. Canter: Thesis (1967), unpublished.
49 R.E. Mischke, et al.: Phys. Rev. Letters 18, 138 (1967).
50 M.Y. Balat, et al.: Phys. Letters 26B, 320 (1968).
51 R.K. Carnegie: Thesis (1968), Princeton TR44.
52 W.A. Mehlhop, et al.: Phys. Rev. 172, 1613 (1968).
53 M. Bott-Bodenhausen, et al.: CERN 69-7, 329.
54 H. Faissner, et al.: Phys. Letters 30B, 204 (1969).
55 M. Cullen, et al.: Phys. Letters 32B, 523 (1970).
56 S.H. Aronson, et al.: Phys. Rev. Letters 25, 1057 (1970).
57 M.Y. Balat, et al.: SINP 13, 53 (1971).
58 R.K. Carnegie, et al.: Phys. Rev. D4, 1 (1971).
59 W. Galbraith, et al.: Phys. Rev. Letters 14, 383 (1965).
60 P. Basile, et al.: in Proc. Balaton Conf. (1965).
61 X. de Bouard, et al.: Phys. Letters 15, 58 (1965).
62 V.W. Fitch, et al.: Phys. Rev. 164, 1711 (1967).
63 R. Messmer, et al.: Phys. Rev. Letters 30, 876 (1973).
64 C. Rubbia: in Proc. 16th Intern. Conf. High Energy Physics, Batavia (1972), p. 157.
65 M. Holder, et al.: Phys. Letters 40B, 141 (1972).
66 M. Banner, et al.: Phys. Rev. Letters 28, 1597 (1972).
67 A. Böhm, et al.: Nucl. Phys. B9, 605 (1969).
68 D.A. Jensen, et al.: Phys. Rev. Letters 23, 615 (1969).
69 M. Bott-Bodenhausen, et al.: Phys. Letters 24B, 194 and 438 (1967).
70 R. Benett, et al.: Phys. Letters 27B, 248 (1968).
71 R. Benett, et al.: Phys. Letters 29B, 317 (1969).
72 R.K. Carnegie, et al.: Phys. Rev. D6, 2335 (1972).
73 J.C. Chollet, et al.: Phys. Letters 31B, 658 (1970).
74 B. Wolff, et al.: Phys. Letters 36B, 517 (1971).
75 G. Barbiellini, et al.: Phys. Letters 43B, 529 (1973).
76 H. Wahl: in Proc. 17th Conf. on High Energy Physics, London (1974);
K. Kleinknecht: Proc. 17th Conf. High Energy Physics, London (1974), Rapporteur's talk.
77 W. Blum, et al.: 4th Intern. Conf. High Meson Spectr., Boston (1974).
78 K. Winter: Proc. Amsterdam Intern. Conf. Elementary Particles (1971), p. 333.

79 W.T. Eadie, et al.: Statistical Methods in Experimental Physics (North-Holland Publishing Co., Amsterdam 1971).
80 J.A. Anderson, et al.: Phys. Rev. Letters 14, 475 (1965); 15, 645 (1965).
81 L. Behr, et al.: Phys. Letters 22, 540 (1966).
82 B.R. Webber, et al.: Phys. Rev. 1D, 1967 (1970).
83 G.W. Meissner, et al.: Phys. Rev. 3D, 59 (1971).
84 Y. Cho, et al.: Phys. Rev. 3D, 1557 (1971).
85 L.H. Jones, et al.: Nuovo Cimento 9A, 151 (1972).
86 M. Metcalf, et al.: Phys. Letters 40B, 703 (1972).
87 M.L. Mallary, et al.: Phys. Rev. D7, 1953 (1973).
88 V.V. Barmin, et al.: Phys. Letters 46B, 465 (1973).
89 S.L. Glashow, S. Weinberg: Phys. Rev. Letters 14, 835 (1965).
90 P. Franzini, et al.: Phys. Rev. 140B, 127 (1965).
91 B. Feldman, et al.: Phys. Rev. 155, 1611 (1967).
92 F. James, H. Briand: Nucl. Phys. B8, 365 (1968).
93 L.S. Littenberg, et al.: Phys. Rev. Letters 22, 654 (1969).
94 Y. Cho, et al.: Phys. Rev. D1, 3031 (1970).
95 F.J. Sciulli, et al.: Phys. Rev. Letters 25, 1214 (1970).
96 P.M. Mautsch, et al.: Preprint COO-1195-210, University of Illinois (1971).
97 B.R. Webber, et al.: Phys. Rev. Letters 21, 498 (1968); Phys. Rev. D3, 64 (1971).
98 M. Baldo-Ceolin, et al.: Paper submitted to Amsterdam Conf. Elementary Particles (1971).
99 G. Neuhofer, et al.: Paper submitted to Amsterdam Conf. Elementary Particles (1971);
F. Niebergall, et al.: Phys. Letters B49, 103 (1974).
100 W.A. Mann, et al.: Phys. Rev. D6, 137 (1972).
101 J.C. Hart, et al.: Nucl. Phys. B66, 317 (1973).
102 O. Fackler, et al.: Phys. Rev. Letters 31, 847 (1973).
103 K.R. Schubert, et al.: Phys. Letters 31B, 662 (1970).
104 F. Graham, et al.: Nuovo Cimento 9A, 166 (1972).
105 J.S. Steinberger: CERN Yellow Report 70-1 (1970).
106 J.M. Gaillard: Proc. Darebury Study Weekend No. 2, DNPL/R9 (1971), p. 161.
107 R. Morse, et al.: Phys. Rev. Letters 28, 388 (1972);
G.V, Dass, P.K. Kabir: Preprint (1972).
108 I. Sandweiss, et al.: Phys. Rev. Letters 30, 1002 (1973).
109 F. Mesei: Z. Physik 255, 146 (1972);
P.D. Miller: Proc. 2nd Intern. School Neutron Physics, Alushka (1974);
see also J. Byrne, et al.: Proposal 1974, University of Sussex.
110 W.B. Dress, et al.: Phys. Rev. D7, 3147 (1973).
111 L.B. Okun, P. Pontecorvo: Zh. Eksper. i. Teor. Fiz. 32, 158 (1957).
112 V. Soergel: Proc. 2nd Intern. Aix-en-Provence Conf. Elementary Particles (1973), C1-84.
113 R.G. Sachs: Phys. Rev. Letters 13, 286 (1964).
114 R. Golub, J.B. Pendlebury: Contemp. Phys. 13, 519 (1972);
L. Wolfenstein: Nucl. Phys. B77, 375 (1974).
115 F. Salzman, G. Salzman: Phys. Letters 15, 91 (1965); Nuovo Cimento 41A, 443 (1965).
116 T.D. Lee: Physics Reports 9C, 143 (1974).
117 R.N. Mohapatra: Phys. Rev. D6, 2023 (1972).
118 A. Pais, J.R. Primack: Phys. Rev. D8, 3036 (1973).
119 J. Frenkel, M.E. Ebel: Nucl. Phys. B83, 177 (1974).
120 N. Mohapatra, J.C. Pati: Phys. Rev. D11, 566 (1975).
121 For a discussion of older milliweak models see L. Wolfenstein: Theory and Particle Phenomenology in Particle Physics (Academic Press 1969), p. 218. For a summary of models derived from gauche-invariant theories see, e.g., WOLFENSTEIN [114].

122 E.E. Overseth, R.F. Roth: Phys. Rev. Letters 19, 321 (1967).
123 W.E. Cleland, et al.: Nucl. Phys. B40, 221 (1972).
124 S.W. Barnes, et al.: Phys. Rev. 117, 238 (1960);
L.D. Roper, et al.: Phys. Rev. 138B, 190 (1965).
125 P. Calaprice, et al.: Phys. Rev. Letters 18, 918 (1967).
126 L. Caneschi, L. Van Hove: CERN 67-27.
127 W. von Witsch, et al.: Phys. Rev. Letters 19, 524 (1967).
128 W.G. Weitkamp, et al.: Phys. Rev. 165, 1233 (1968).
129 N. Cabbibo: Phys. Rev. Letters 14, 965 (1964).
130 I. Prentki, V. Veltman: Phys. Letters 15, 88 (1965).
131 A. Pais: Phys. Rev. Letters 3, 242 (1959).
132 C. Baltay, et al.: Phys. Rev. Letters 15, 591 (1965).
133 O. Dobrzynski, et al.: Phys. Letters 22, 105 (1966).
134 H. Rollnik: Notes on theoretical seminar on $K^0$ decays at Bonn in the winter semester 1973/74.

## Licht und Materie 1b
## Light and Matter 1b

Editor: L. Genzel

34 figures. XVI, 538 pages. 1974 (Handbuch der Physik, Band 25, Teil 2 b)

*Contents:* Birman, J. L., Theory of Crystal Space Groups and Infra-Red and Raman Lattice Processes of Insulating Crystals: Scope and plan of the article. The crystal space group. Irreducible representations and vector spaces for finite groups. Irreducible representations of the crystal translation group $\chi$. Irreducible representations and vector spaces of space groups. Reduction coefficients for space groups: Full group methods. Reduction coefficients for space groups: Subgroup methods. Space group theory and classical lattice dynamics. Space-time symmetry and classical lattice dynamics. Applications of results on symmetry adapted eigenvectors in classical lattica dynamics. Space-time symmetry and quantum lattice dynamics. Interaction of radiation and matter: Infra-red absorption and Raman scattering by phonons. Group theory of diamond and rocksalt space groups. Phonon symmetry, infra-red absorption and Raman scattering in diamond and rocksalt space groups. Some aspects of the optical properties of crystals with broken symmetry: Point imperfections and external stresses. Respicc, adspice, prospice. Acknowledgements.

## Light Scattering in Solids

Editor: M. Cardona. 111 figures, 3 tables. XIII, 339 pages. 1975

(Topics in Applied Physics, Vol. 8)

*Contents:* M. Cardona, Introduction. — A. Pinczuk; E. Burstein, Fundamentals of Inelastic Light Scattering in Semiconductors and Insulators. — R. M. Martin; L. M. Falicov, Resonant Raman Scattering. — M. V. Klein, Electronic Raman Scattering. — M. H. Brodsky, Raman Scattering in Amorphous Semiconductors. — A. S. Pine, Brillouin Scattering in Semiconductors. — Y.-R. Shen, Stimulated Raman Scattering.

## Light Scattering Spectra of Solids

Proceedings of the International Conference, 1968 at New York University

Editor: G. B. Wright. 282 figures. XIX, 763 pages. 1969

*Contents:* Introductory Remarks. — Phonons and Polaritons. — Phonons. — Magnons and Other Electronic Excitations. — Free Carriers. — Phonons, Resonance Scattering, Metals, Morphic Effects. — Mixed Crystals and Point Defects. — Brillouin Scattering. — Phase Transitions and Critical Scattering. — List of Participants. — Author Index. — Topical References.